FIREFIGHTER WRITTEN PRACTICE EXAMS

VOLUME #1

1993

BY

ARTHUR R. COUVILLON

FIRST EDITION

Printed in the United States of America

Library of Congress Cataloging Data:

Couvillon, Arthur R.
"Firefighter Written Practice Exams"
1. Fire Service Entrance Exam Preparation, text books, etc.
2. Fire Service: Handbooks, manuals, text books, etc.
3. Fire Engineer Handbooks, manuals, text books, etc.
4. Fire Captain: Handbooks, manuals, text books, etc.
5. Firefighter: Handbooks, manuals, text books, etc.
6. Information: Handbooks, manuals, text books, etc.
7. Promotional: Handbooks, manuals, text books, etc.
8. Practice Exams: Fire Service
I. Title
ISBN 0-938329-64-2
LCCN 93-077195

INTRODUCTION

The intent of this book is to provide exam practice pertaining to the knowledge and duties necessary for the position of: **FIREFIGHTER.**

The goal of this book is to help provide the knowledge to any motivated individual desiring to secure the position of: **FIREFIGHTER.**

A vast number of Cities across the U.S.A. give exams for the position of: **FIREFIGHTER.** The exams included in this book are presented so as to cause studying to be more organized and easier. As you read through this book you will come across some duplication of information, this information will be presented in a slightly different manner, as a study technique, which may help some individuals to retain certain bits of information.

Excluding information of a local nature and particular department policy, you will find that the study material covered in this book will be an excellent resource for you to use in studying for **ENTRANCE LEVEL FIRE DEPARTMENT WRITTEN EXAMS.** Included are enormous amounts of required knowledge and information that each individual will need in order to pass and increase the final score on the exam.

Remember most all Fire Department exams will have questions that will cover their own State and local laws, codes and ordinances along with the departments policy and procedures, rules and regulations, etc. Make sure that you cover these particulars for the department you are testing for!

ACKNOWLEDGEMENTS

Appreciation is expressed to those individuals, Rookie Firefighters, Fire Engineers, Fire Captains, Fire Training Officers, Fire Technology Instructors, Fire Chiefs, and Personnel Directors that have contributed to the compilation of this book: "FIREFIGHTER WRITTEN PRACTICE EXAMS".

ABOUT THE AUTHOR

Arthur R. Couvillon is a veteran Firefighter, with over 20 years of Firefighting experience. Art shares the knowledge that he has gained during this period along with research and surveys that he has taken with fire authorities all across the U.S.A. Listed below are some of Art's published Fire Service Study Books:

THE COMPLETE FIREFIGHTER CANDIDATE

FIREFIGHTER WRITTEN EXAM STUDY GUIDE

FIREFIGHTER ORAL EXAM STUDY GUIDE

FIRE SERVICE ENTRANCE EXAM PREPARATION, Instructor Manual & Student Manual

ADVANCEMENT IN THE FIRE SERVICE

FIRE ENGINEER WRITTEN EXAM STUDY GUIDE

FIRE ENGINEER ORAL EXAM STUDY GUIDE

FIRE SERVICE HYDRAULICS

FIRE CAPTAIN WRITTEN EXAM STUDY GUIDE

FIRE CAPTAIN ORAL EXAM STUDY GUIDE

FIREFIGHTER WRITTEN PRACTICE EXAMS

FIRE ENGINEER WRITTEN PRACTICE EXAMS

FIRE CAPTAIN WRITTEN PRACTICE EXAMS

All books are available from:
INFORMATION GUIDES
P.O. Box 531, Hermosa Beach, CA 90254

1 - 800 - "FIRE - BKS"

TABLE OF CONTENTS

PRACTICE EXAM #1

FIREFIGHTER WRITTEN
PRACTICE EXAM #1

1. Certain fire gases are a hazard to human life. Which one of the following ranks first in the cause of fire deaths:

   A. nitrogen
   B. carbon monoxide
   C. hydrogen
   D. carbon dioxide

2. Carbon monoxide gas is best characterized as:

   A. pink colored
   B. having a strong odor
   C. tasting like bitter almonds
   D. slightly lighter than air

3. Of the following disorders, which one would display: fever, chills, inflamed eyelids, running nose, and a cough:

   A. chicken pox
   B. tuberculosis
   C. measles
   D. scarlet fever

4. A respirator is used by firefighters:

   A. to extinguish small blazes
   B. to demolish burning buildings
   C. when necessary to ventilate
   D. to save lives
   E. to control a conflagration

5. The object shown below is used to:

   A. countersink
   B. back screws out
   C. remove rivets
   D. drill in glass

6. Of the following two drawings, which one has the weights that will move?

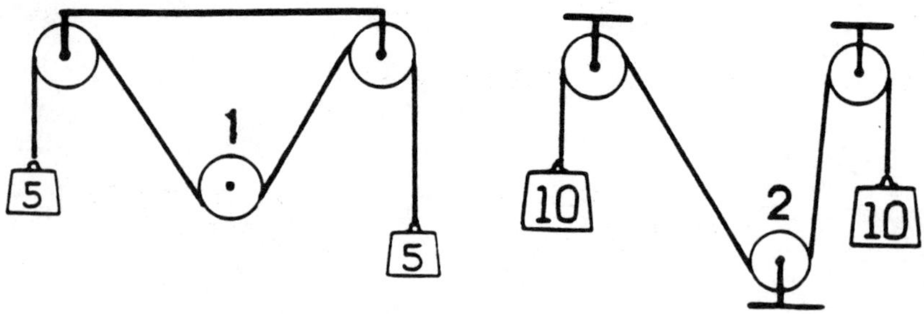

GROUND LEVEL

A. drawing #1
B. drawing #2

7. Of the four drawings, which one has the weights that will move?

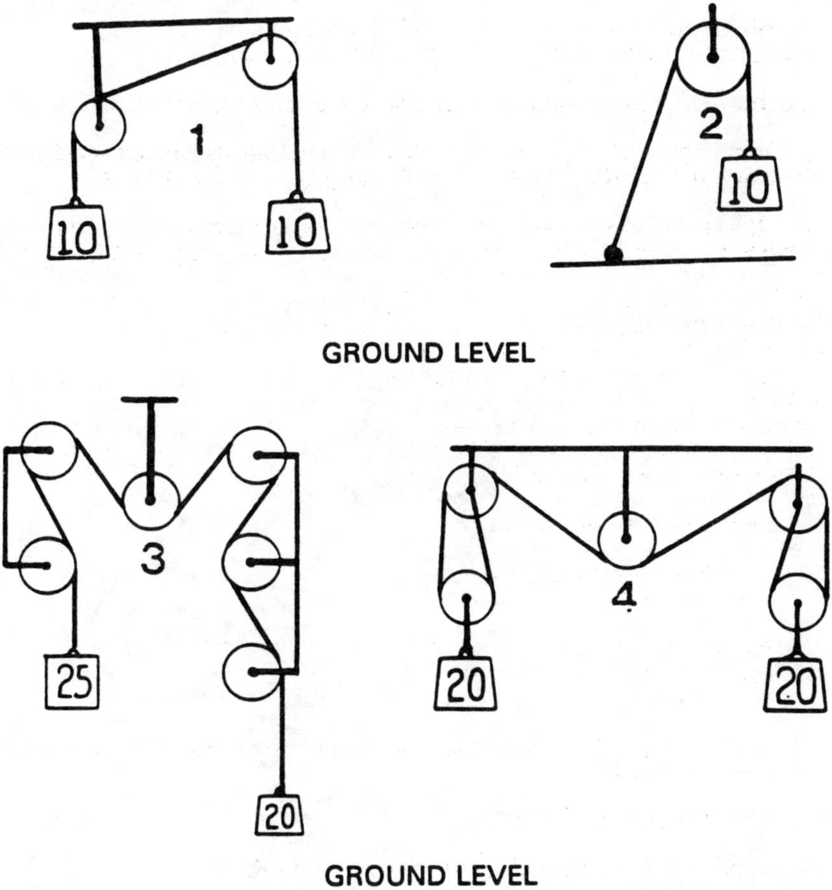

A. drawing #1
B. drawing #2
C. drawing #3
D. drawing #4

8. In the drawing below, when the top pulley rotates in the direction of the arrow, in what direction will the lower pulley rotate?

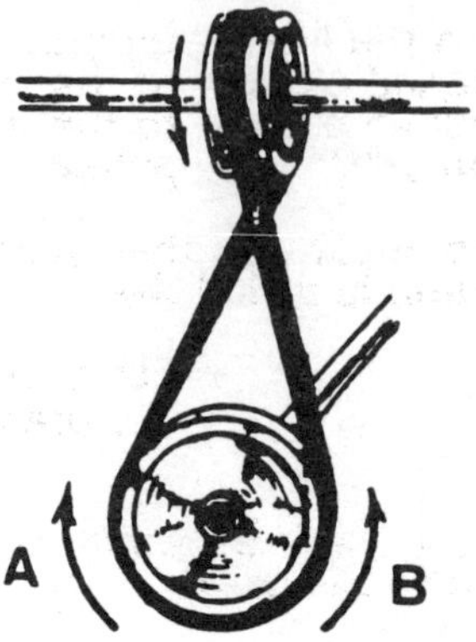

A. direction A
B. direction B
C. both directions

9. In the drawing below:

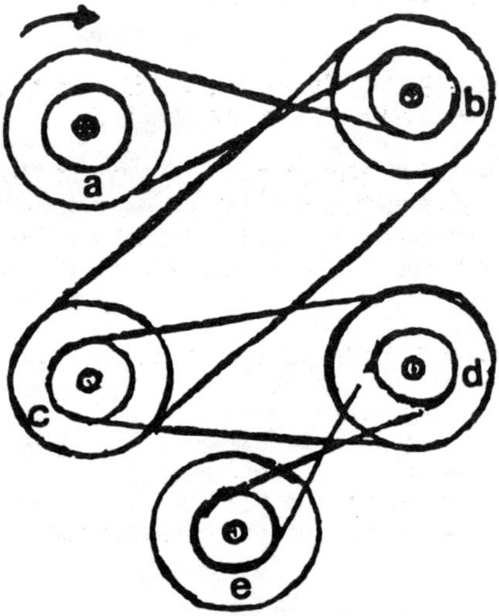

A. wheels a and d turn in the opposite direction
B. wheel d turns faster than wheel b
C. wheel a turns faster than wheel c
D. wheels d and a turn at the same rate

10. Room temperature is normally higher near the ceiling of a room than near the floor because:

A. hot water pipes are near the ceiling
B. the warmer air is lighter than the cooler air
C. the air circulates better at floor level
D. most openings are nearer the floor

11. The proper use of the tension pulley on a belt that connects two different size pulleys:

A. will not add to the life of the pulley
B. usually positioned closest to the smaller pulley
C. usually positioned closest to the larger pulley
D. usually will be a "V" type pulley

12. The deice on fire apparatus that indicates the revolutions per minute of the pump is called the:

A. barometer
B. speedometer
C. ammeter
D. tachometer
E. pressure gauge

13. If the temperature gage in a fire pumper indicates that the engine is starting to overheat:

A. add water to radiator
B. pour hot water in immediately
C. pour in anti-freeze immediately
D. allow it to cool down

14. The reason alcohol is added to the radiator of a fire apparatus in cold weather is because it:

A. lowers the freezing point of the mixture
B. lowers the boiling point of the mixture
C. raises the freezing point of the mixture
D. raises the boiling point of the mixture

15. When a mass of air rises, it is cooled primarily because:

A. it expands
B. moisture condenses
C. ice crystals are formed
D. it mixes with cold air
E. it is closer to cold outer space

16. Of the following, the most inept conductor of electricity is:

A. iron
B. steel
C. graphite
D. rubber
E. copper

17. The one among the following which is the name of a commonly used roller bearing is:

A. Alemite
B. Hotchkiss
C. Stewart
D. Timkin
E. Vacuum

18. A fire can cause structural damage to concrete and is discernable when the concrete shows signs of:

A. cracking B. calcination C.discoloration D. all of the above

19. Static pressure:

A. is the pressure created by the weight of water in motion.
B. depends upon the length and shape of the container
C. is usually stated in pounds per square foot
D. depends upon the vertical distance

20. The watt is a unit of electrical:

A. current B. power C. potential D. inductance

21. Of the following, the best conductor of heat is:

A. water B. steel C. copper D. glass

22. The freezing point on a Fahrenheit thermometer is____degrees:

A. 0 B. 10 C. 28 D. 32

23. Natural gas consists primarily of:

A. propane B. methane C. ethane D. butane

24. Greater uniformity has been attained in laws for arson control among the various states then for any other fire prevention feature. On the basis of this statement it would be most reasonable to assume that arson control:

A. has been incorporated into the fire code of all the states
B. is intimately related to fire prevention
C. laws are the same for all states
D. is perhaps the outstanding problem with which state legislators have to contend

25. A deluge sprinkler system may be described as a system of open sprinkler heads, or a combination of open and closed sprinklers, controlled by a quick opening mechanical or hydraulic valve, the latter being caused to operate, either automatically, as by a thermostat system or manually. Considering the imply of the preceding statement, it can most precisely be indicated that the deluge sprinkler:

A. is operated on the principal of hydraulic pressure
B. may be controlled by a hydraulic valve which can be operated by hand
C. hydraulically controlled is far more effective than one mechanically controlled
D. is installed to insure adequate protection by the sprinkling process

26. The primary basis for insulating pipes in heating is to:

A. prevent the loss of heat
B. prevent pipes from rusting
C. prevent fires
D. prevent injury to people

27. Some fires are caused by arcing. Firefighters should be aware that this term means:

A. an electric current flowing along a conductor bent into a curve
B. an electric current passing across a gap between two conductors
C. the blowing out of an electric fuse
D. a trolley wire coming into contact with the round rail

28. The smallest division that a substance may be divided into without destroying the substance is:

A. atom B. particle C. nucleus D. molecule

29. If a pile of old clothes and rubbish catches fire in a corner, the most likely cause of the fire would be:

A. arson
B. faulty wiring in the building
C. sunlight focused on the pile through a window
D. spontaneous combustion
E. faulty heating equipment

30. Substances that are incapable of combustion are:

A. flammable
B. fireproof
C. fire hard
D. incinerated

31. The earth rotates through 360 degrees of longitude in 24 hours. How many minutes does it take to revolve through 1 degree:

A. .25
B. 6
C. 15
D. 25
E. none of the above

32. A fire science instructor bought 3 gross of pencils at $.38 a dozen. He sold the pencils at $.05 each. How much was his profit per gross:

A. $7.92
B. $2.64
C. $0.22
D. $0.66
E. none of the above

33. Add: 13 + 45 + 29 =

A. 77
B. 86
C. 88
D. 97
E. none of the above

34. Add: 97 + 48 + 56 =

A. 182
B. 201
C. 203
D. 211
E. none of the above

35. Add: 63 + 58 + 28 =

A. 139
B. 146
C. 149
D. 150
E. none of the above

36. Subtract: 93 - 64 =

A. 28
B. 29
C. 31
D. 38
E. none of the above

37. Subtract: 116 - 47 =

A. 59
B. 69
C. 71
D. 79
E. none of the above

38. Multiply: (46)(7)

A. 282
B. 312
C. 316
D. 322
E. none of the above

39. How many feet of rope are needed to secure a protective area around a hazardous structure that is 25ft wide and 30ft long?

A. 100 feet B. 110 feet C. 120 feet D. 130 feet

40. 1/2 of 1/4 is:

A. 1/12 B. 1/8 C. 1/2 D. 1/4 E. .8

41. Firefighter Charron bought a pair of boots at a 40% discount. The boots cost $165.00. What was the original price of the boots?

A. $66
B. $231
C. $275
D. $412.5
E. none of the above

42. On a $9840 bill for fire equipment, what is the difference between a discount of 30% and a discount series of 20% and 10%?

A. no difference
B. $196.80
C. $787.20
D. $2755.20

43. How many cubic yards of dirt had to be removed to make an trench 30 feet long, 21 feet wide and 6 feet deep:

A. 1260
B. 3780
C. 140
D. 420
E. none of the above

44. Which of the "products of combustion" in a fire is predominately responsible for the extension of the fire:

A. heat
B. smoke
C. flame
D. fire gases

45. A reduction of oxygen content of the air from the normal 21% to 15% will extinguish most fires in places which do not involve substances which will produce glowing embers or smoldering fire." Judging from this information alone, it follows:

A. when the oxygen content of the air is negligible, fire cannot long be sustained
B. a simple reduction of 6% in the oxygen content of the air will extinguish the greater majority of fires
C. a smoldering fire cannot be fought by reducing the oxygen content of the air
D. water is the best extinguishing agent in the case of a glowing ember fire

46. Analysis has shown that the stresses in a beam from the bending movements are:

A. greatest at the top and bottom of the beam
B. greatest at the bottom only
C. greatest in the center
D. evenly distributed throughout the beam

47. During events in an auditorium/theater containing scenery and/or equipment-machinery, routinely a firefighter will be assigned to the event. The primary basis for this assignment is that:

A. theaters are located in districts if high property value
B. the use of scenery and machinery increases the fire hazard.
C. theatrical districts have heavy traffic, making for slow response of apparatus.
D. emergency exits may be blocked by the scenery or machinery.

48. Firefighters should be:

A. dedicated
B. loyal to the department
C. faithful to fellow firefighters
D. all of the above

49. As a probationary firefighter, you get an idea for improving apparatus maintenance and you mention it to a senior firefighter. At the next company meeting, your captain openly compliments the senior firefighter for his superb recommendation, which was your idea. The action you should take in this situation is to:

A. tell the other members of the company the whole story after the inspection
B. Ask for advice from another older member
C. forget about the incident since this man will probably be helpful to you in return
D. do nothing about it, but next time make your suggestion to your superior officer

50. The primary purpose of an air supply unit is to:

A. supply SCBA equipment
B. supply air for pneumatic tools
C. ventilate smoke filled buildings
D. disperse gaseous hazardous materials

51. Following the extinguishment of a fire, one firefighter usually stays at the scene after the fire companies leave, of the following, the primary basis for this procedure is that the firefighter can:

A. prevent looters from stealing valuables
B. watch for any rekindling of the fire
C. search the area for lost valuables
D. examine the premises for evidence of arson

52. One week after June 3rd is:

A. June 4
B. June 9
C. June 10
D. June 13

53. Fire departments invariably attempt to obtain the finest apparatus and equipment and maintain them in the best repair. The principal reason for this policy is that:

A. public confidence in the department is increased
B. failure of equipment at a fire may have serious consequences
C. replacement of worn out parts is often difficult
D. the dollar cost to the department is less in the long run

54. Of the many unknown factors confronting the firefighter arriving at the scene of a building involved in fire, the one factor that can be changed with the greatest degree of accuracy is the:

A. building occupancy
B. building itself
C. life hazard
D. necessary fire flow

55. Of the choices below, the primary reason for utilizing a small nozzle and hose line for an interior apartment fire:

A. They are more efficient
B. large hose is too bulky
C. they are put in service faster
D. they will create less water damage

56. While in the fire academy, your group assists at a fire. After the fire is under control, an older firefighter, who has no authority over you, discloses to you that he was observing your performance. He recommends specific changes in your technique. From choices below, the best course of action is to:

A. thank him for his advice and tell him you will use it when you find yourself in difficulty
B. discuss the changes he proposes with him and then take the action which seems best to you
C. thank him for his advice and bring up his suggestions at the next class session
D. listen to him, thank him courteously, but ignore his suggestion

57. Assume the owner of a burning apartment building complains that although the fire is situated on the ground floor, firefighters are cutting ventilation holes in the roof. Of the choices below, the primary reason that can be given for their action is that the fire can be fought most effectively by allowing:

A. smoke and hot gasses to escape
B. firefighters to attack the fire from above
C. firefighters to gain access to the building through the holes
D. immediate inspection of the roof area for fire extension

58. Which of the following is LEAST identified with firefighting:

A. chemistry B. hydraulics C. mechanics D. geology

59. Of the choices below, what is the predominant reason for fire hydrants being broken:

A. hydrants are usually unprotected
B. drivers usually cannot see fire hydrants
C. replacement cost of fire hydrants is not very high
D. hydrants are usually located in the center of the block and they are unprotected

60. Atmospheric air contains___% oxygen?

A. 21 B. 15 C. 12 D. 34

61. Hot magnesium reacting with water liberates:

A. magnesium chloride
B. nitrous oxide
C. hydrogen
D. nitrogen

62. An electron is a/an:

A. particle of negative electricity
B. static charge
C. hydrogen atom
D. electronic nucleus

IN THE FOLLOWING NUMBER PROGRESSIONS, CHOOSE THE NUMBER WHICH LOGICALLY COMES NEXT IN THE SERIES (questions #63 - #65)

63.
1 3 5 7 9 11 13 ___

A. 15 B. 11 C. 18 D. 23

64.
2 4 8 16 32 64 128 ___

A. 64 B. 96 C. 128 D. 256

65.
20 10 12 6 4 ___

A. 2 B. 12 C. 10 D. 24

INSERT THE CORRECT LETTER THAT WILL COMPLETE THE FOLLOWING ALPHABETICAL PROGRESSIONS(questions #66-#68)

66.
V T R P __

A. O B. N C. X D. Z

67.
CC EE GG II __

A. BB B. FF C. KK D. JJ

68.
A C F J _

A. M B. P C. O D. Q

69. How many blocks are there in the drawing below?

A. 7
B. 8
C. 9
D. 10

70. How many blocks are thee in the drawing below?

A. 25
B. 33
C. 35
D. 36

IN THE FOLLOWING DRAWINGS, FIND THE CUBE AT THE RIGHT THAT COULD BE THE ORIGINAL CUBE ROTATED INTO A DIFFERENT POSITION. CUBES MAY BE ROTATED INTO ANY POSITION AND THERE IS A DIFFERENT DESIGN ON EACH OF THE SIX SURFACES OF EACH CUBE. (questions #71 & #72)

71.

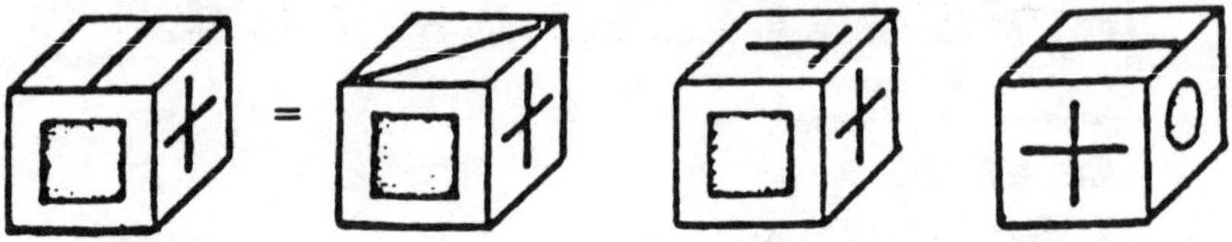

ORIGINAL A. B. C.

72.

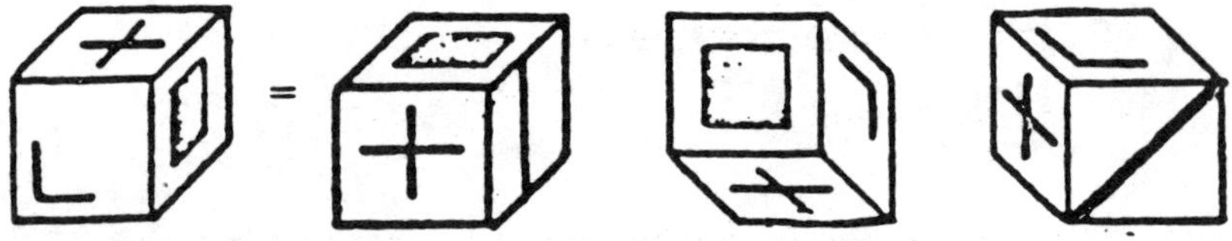

ORIGINAL A. B. C.

IN THE FOLLOWING DRAWINGS DETERMINE THE RELATIONSHIP BETWEEN THE FIGURES LOCATED IN THE TOP TWO BOXES AT THE LEFT, THEN DETERMINE HOW THE FIGURE IN THE BOTTOM LEFT BOX IS SOMEHOW RELATED TO ONE OF THE FOUR FIGURES TO THE RIGHT: REMEMBER POSITION AND DIRECTION. (#73-#75)

73.

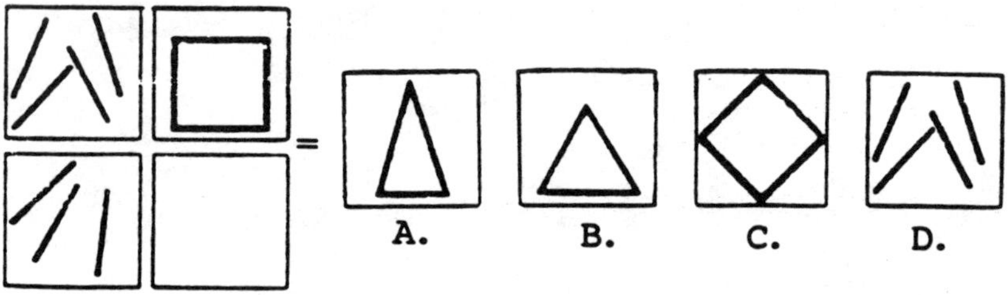

A. B. C. D.

74.

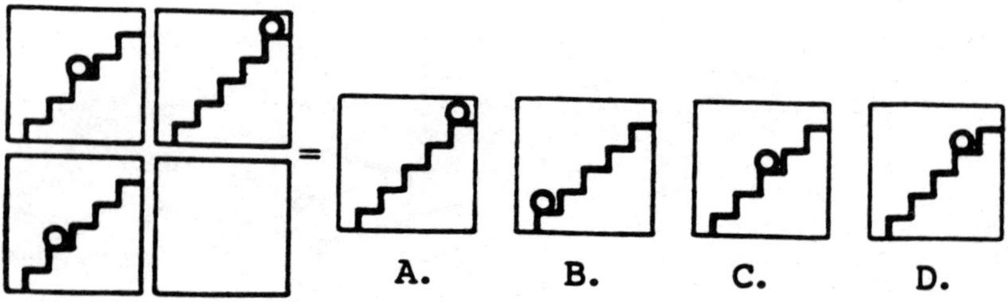

A. B. C. D.

75. 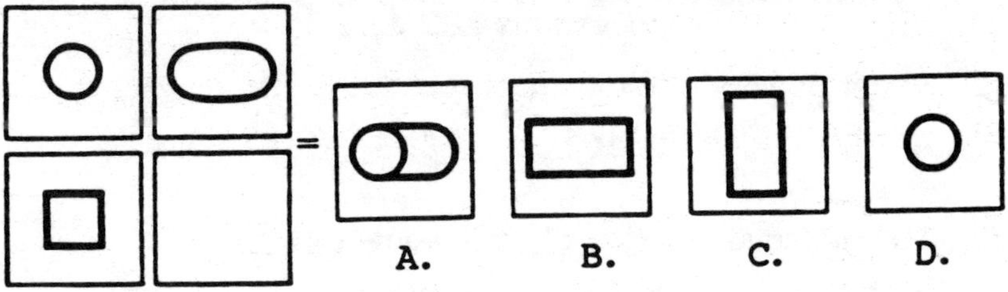

76. Narrow means most nearly:

A. attractive
B. not wide
C. rough
D. without trees

77. Defective means most nearly

A. old-fashioned
B. uneven
C. unused
D. faulty

78. Caution means most nearly:

A. care
B. fear
C. danger
D. difficulty

79. Portable means most nearly:

A. business-like
B. practical
C. difficult use
D. easily transported
E. of foreign manufacture

80. Obsolete means most nearly:

A. out-of-date
B. broken down
C. good as new
D. replaced
E. remolded

81. Fluctuate means most nearly:

A. vary
B. divide
C. improve
D. irritate
E. undertake

STATEMENT

Vapor density equals the weight of a vapor-air mixture resulting from the vaporization of a flammable liquid at equilibrium temperature and pressure conditions, as compared with the weight of an equal volume of air under the same conditions. Vapor density is the relative density of vapor gas, with no air present, compared to air. Vapor density is a ratio of gases. The vapor density of air is equal to 1, gases of less than 1 indicates that the gas is lighter than air. In order to compare vapor densities of gas, you must know that the molecular weight of air is equal to 29.

ACCORDING TO THE PREVIOUS STATEMENT:
(questions #82 & #83)

82. Vapor density of air is equal to:

A. 15 B. 10 C. 5 D. 1

83. The molecular weight of air is equal to:

A. 1 B. 11 C. 19 D. 29

STATEMENT

There is hardly a city in the country that is not short of fire protection in some areas within its boundaries. These municipalities have spread out and have reshuffled there residential, business, and industrial districts without readjusting the existing protective fire forces, or creating new protection units. Fire stations are still situated according to the needs of earlier times and have not been altered or improved to house modern firefighting equipment. They are neither efficient for carrying out their tasks nor livable for the men who must occupy them.

ACCORDING TO THE PREVIOUS STATEMENT
(questions #84 - #88)

84. Of the following, the title which best describes the central idea of the previous statement:

A. The Energetic Nature of Modern Society
B. The Price of Fire Protection
C. The Location and Design of Fire Stations
D. The Purpose and Use of Fire Fighting Apparatus

85. According to the statement above, fire protection is inadequate in the United States in:

A. most areas of some cities
B. some areas of most cities
C. some areas of all cities
D. all areas of some cities
E. most areas of most cities

86. The one of the following criteria for planning of fire stations which is not mentioned in the statement above:

A. comfort of firefighters
B. appropriate location
C. design for modern equipment
D. price of construction

87. Of the following recommendations for improving the fire service, the one which would best manage the difficulty characterized in the statement above would involve:

A. specialized instruction in the use of contemporary fire apparatus
B. replacement of antiquated fire apparatus
C. extended primary training for probationary firefighters
D. reassignment of fire districts

88. The tone of the writer of the statement above may best be illustrated by which one of the following adjectives:

A. encouraging
B. pessimistic
C. struggle
D. critical

THE QUESTIONS NUMBERED 89-94, INCLUSIVE, CONSIST OF SENTENCES WHICH MAY BE CORRECT OR WHICH MAY CONTAIN ERRORS IN SPELLING, GRAMMAR, OR PUNCTUATION. IF THE SENTENCE IS CORRECT, MARK A. IF THE SENTENCE CONTAINS A SPELLING ERROR, MARK B. IF THE SENTENCE CONTAINS AN ERROR IN GRAMMAR, MARK C. IF THE SENTENCE CONTAINS AN ERROR IN PUNCTUATION, MARK D.

89. The performance of the apparatus, was evaluated by qualified engineers conducting exhaustive analyses.

A B C D

90. The importance of the department's operations is not appreciated sufficiently by the legislative committee that controls operations.

A B C D

91. Those who have the opportunity to observe and study the shifting scene in the nation's fire service had been aware of the necessity for more dynamic leadership.

A B C D

92. The apprentice accidently turned the wrong knob; a violent explosion resulted.

A B C D

93. Despite the heroic efforts of the volunteer firemen, the conflagration raged out of control, consuming everything in its destructive path.

A B C D

94. On this melancholy occasion, the minister said, "we must dedicate ourselves anew to the prevention of wanton destruction of lives and property by fire."

A B C D

95. Which of the following words is misspelled?

A. confidence
B. maturity
C. aspirations
D. insurence
E. none of the above

THE FOLLOWING SENTENCES ARE PART OF A MEANING PARAGRAPH BUT THEY ARE NOT GIVEN IN THEIR PROPER ORDER. YOU ARE TO DECIDE WHAT WOULD BE THE BEST ORDER IN WHICH TO PUT THE SENTENCES SO AS TO FORM A WELL-ORGANIZED PARAGRAPH. EACH SENTENCE HAS A PLACE IN THE PARAGRAPH;THERE ARE NO EXTRA SENTENCES. YOU ARE THEN TO ANSWER QUESTIONS 96-98 INCLUSIVE ON THE BASIS OF YOUR REARRANGEMENT OF THE SCRAMBLED SENTENCES INTO A PROPERLY ORGANIZED PARAGRAPH.

In 1887 some insurance companies organized an Inspection Department to advise their clients on all phases of fire prevention and protection. Probably this has been due to the smaller annual fire losses in Great Britain than the United States. It tests various fire prevention devises and appliances and determines manufacturing hazards and their safeguards. Fire research began earlier in the United States and is more advanced than in Great Britain. Later they established a laboratory specializing in electrical, mechanical, hydraulic and chemical fields.

96. When the five sentences above are arranged in the proper order, the paragraph starts with the sentence which begins:

A. "In 1887..."
B. "Probably this..."
C. "It tests..."
D. "Fire research..."
E. "Later they..."

97. In the last sentence listed above, "they" refers to:

A. insurance companies
B. the United States and Great Britain
C. the Inspection Department
D. clients
E. technicians

98. When the above paragraph is properly arranged, it ends with the words:

A. "...and protection."
B. "...the United States."
C. "...their safeguards."
D. "...in Great Britain."
E. "...chemical fields."

99. Which sentence is grammatically correct?

A. Certainly, even the most experienced firefighter feels stress.
B. Certainly even the most experienced firefighter feels stress.
C. Certainly; even the most experienced firefighter feels stress.
D. Certainly: even the most experienced firefighter feels stress.

100. Which sentence is grammatically correct?

A. We rarely eat meat at our firehouse. My captain being a vegetarian.
B. We rarely eat meat at our firehouse my captain being a vegetarian.
C. We rarely eat meat at our firehouse; my captain being a vegetarian.
D. We rarely eat meat at our firehouse, my captain being a vegetarian.

PRACTICE EXAM #1
ANSWERS

1. B
2. D
3. C
4. D
5. A
6. A
7. C
8. A
9. A
10. B
11. B
12. D
13. D
14. A
15. A
16. D
17. D
18. D
19. D
20. B
21. C
22. D
23. B
24. B
25. B
26. A
27. B
28. D
29. D
30. B
31. E
32. B
33. E
34. B
35. C
36. B
37. B
38. D
39. B
40. B
41. C
42. B
43. C
44. A
45. A
46. A
47. B
48. D
49. D
50. A
51. B
52. A
53. B
54. B
55. D
56. C
57. A
58. D
59. A
60. A
61. A
62. A
63. A
64. D
65. A
66. B
67. C
68. C
69. B
70. B
71. C
72. B
73. B
74. D
75. B
76. B
77. d
78. A
79. D
80. A
81. A
82. D
83. D
84. C
85. B
86. D
87. D
88. D
89. D
90. A
91. C
92. B
93. B
94. D
95. D
96. D
97. A
98. C
99. A
100. D

PRACTICE EXAM #2

FIREFIGHTER WRITTEN
PRACTICE EXAM #2

1. Any substance which offers a very high or very great resistance to the passage of electric current through it is called a/an:

   A. conductor B. closed circuit C. fuse D. insulator

2. Water is a better conductor of electricity when:

   A. it contains a large amount of organic matter
   B. it is absolutely pure
   C. its mineral content is high
   D. its mineral content is low

3. A gallon of water weighs about:

   A. 2 1/2 lbs
   B. 4 1/4 lbs
   C. 6 2/3 lbs
   D. 8 1/3 lbs
   E. 10 1/2 lbs.

4. Of the choices below, it is most dangerous for a rope which is to be used in firefighting to come in contact with:

   A. acid
   B. kerosene
   C. linseed oil
   D. shellac
   E. water

5. Ice formation in water pipes often causes the bursting of the pipes because:

   A. the additional weight of the ice overloads the pipes
   B. water cannot pass the ice block and builds up great pressure on the pipes
   C. the cold causes contraction of the pipes and causes them to pull apart
   D. water expands upon freezing and builds up great pressure on the pipes

6. Electrons are:

   A. neutral particles
   B. the nuclei of atoms
   C. positive particles
   D. negative particles

7. The smallest unit that a substance can be divided into without destroying the substance is:

A. atom
B. particle
C. nucleus
D. molecule

8. When radioactive materials are involved with fires, the radiation coming from the fire:

A. increases in rate and intensity
B. decreases in rate and intensity
C. increases in rate and decreases in intensity
D. is not effective

9. Which science does hydraulics pertain to:

A. water/liquids under pressure
B. water/liquids in motion
C. gravity of water
D. water for firefighting

10. Of the following flame colors, which one would indicate the highest temperature:

A. orange-red
B. light red
C. orange-yellow
D. yellow-white

11. Divide 7 into 399

A. 91 B. 93 C. 96 D. 98

12. Divide 90 into 76,500

A. 840 B. 850 C. 880 D. 890

13. Add:

$$\begin{array}{r} 21 \\ 19 \\ 24 \\ 99 \\ \underline{88} \end{array}$$

A. 240 B. 250 C. 261 D. 251

14. Subtract:

$$\begin{array}{r} 68,412 \\ \underline{15,489} \end{array}$$

A. 53,024 B. 52,923 C. 52,823 D. 52,933

15. Add:

$$\begin{array}{r} 6,133 \\ 4,210 \\ 9,641 \\ \underline{7,589} \end{array}$$

A. 26,573 B. 27,573 C. 25,563 D. 27,473

16. The diameter of a wheel is 14 ft. How many times must the wheel turn to travel 440 ft. (use pi=22/7)

A. 6 times B. 8 times C. 10 times D. 12 times

17. A drawing of a certain building is 10 in. by 15 in. On this drawing, 1 in. represents 5 feet. If the same drawing had been made 20 in. by 30 in., 1 in. on the drawing would represent:

A. 7 1/2 ft B. 5 ft C. 10 ft D. 2 1/2 ft

18. A 5 quart solution of sulfuric acid and water is 60% acid. A gallon of water is added. What percent of the resulting solution is acid:

A. 33 1/3%
B. 40%
C. 48%
D. 50%
E. none of the above

19. Firefighter Derouin can polish Engine #81 in twenty minutes and Frank can polish the same apparatus in thirty minutes. How long will it take them working together to polish the apparatus:

A. 10 minutes
B. 12 1/2 minutes
C. 15 minutes
D. 25 minutes
E. none of the above

20. What is the radius of the largest circular disc that can be cut from a strip of metal 15" X 21":

A. 7 1/2"
B. 15"
C. 157 1/2"
D. 315"
E. none of the above

IN THE FOLLOWING DRAWINGS, FIND THE CUBE AT THE RIGHT THAT COULD BE THE ORIGINAL CUBE ROTATED INTO A DIFFERENT POSITION. CUBES MAY BE ROTATED INTO ANY POSITION AND THERE IS A DIFFERENT DESIGN ON EACH OF THE SIX SURFACES OF EACH CUBE (questions #21 - #25)

21.

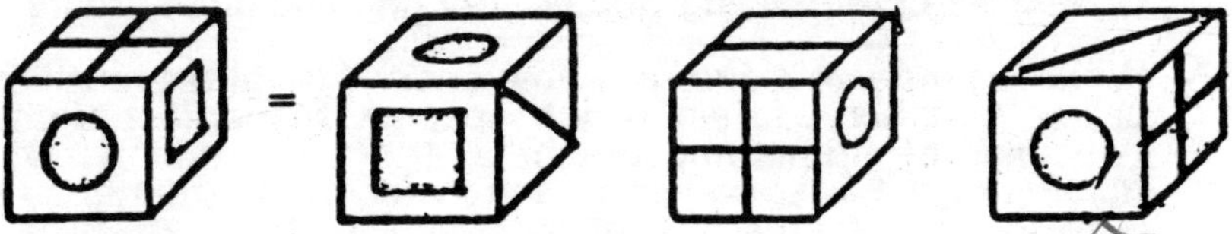

ORIGINAL A. B. C.

22.

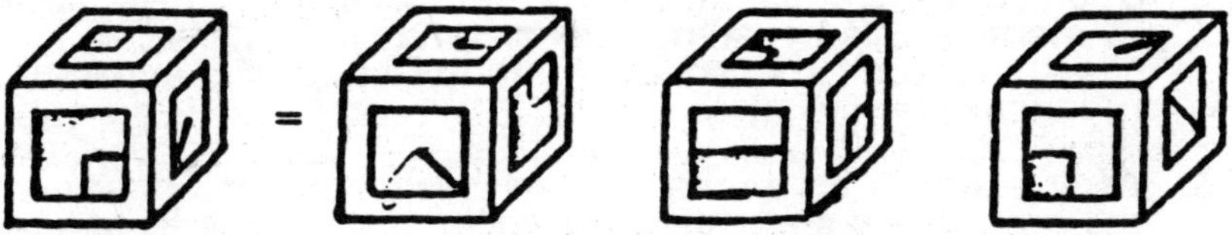

ORIGINAL A. B. C.

23.

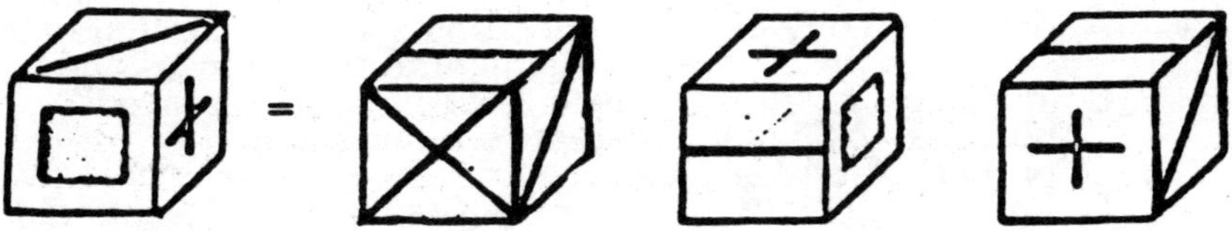

ORIGINAL A. B. C.

24.

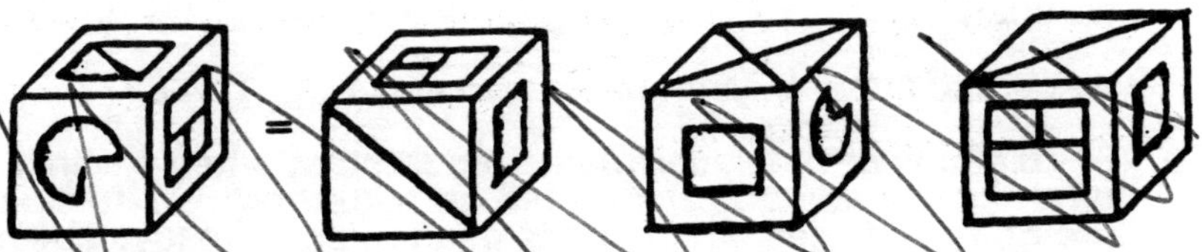

ORIGINAL A. B. C.

25.

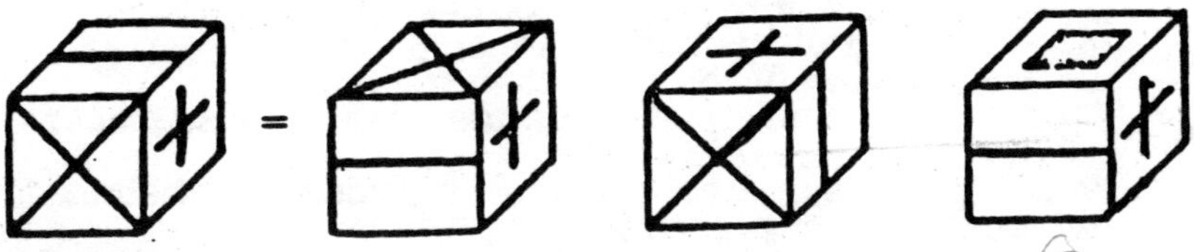

ORIGINAL A. B. C.

26. If a commanding officer gives you certain orders at a fire which seem to you to be ill-advised, generally speaking you should:

A. obey them
B. ask an older firefighter what to do
C. refuse to comply with the order
D. ask the commanding officer why he gave such orders

27. While on duty at a fire, a probationary firefighter receives an order from his Captain which appears to conflict with the principles of firefighting taught at the fire academy. Of the following, the best course of action for the firefighter to take is to follow the order and at a convenient time, after the fire, to:

A. discuss the apparent inconsistency with his Captain
B. discuss the apparent inconsistency with another Captain
C. mention this apparent inconsistency informally
D. ask a more experienced firefighter about the apparent inconsistency

28. After a fire in an apartment has been brought under control, the firefighters are engaged in extinguishing the last traces of the fire. A fireman who notices an expensive vas in the room in which the activities were concentrated moves it to an empty closet in another room. The fireman's action is:

A. proper, chiefly because the owners will realize that extreme care was taken to avoid damage to their possessions
B. improper, chiefly because disturbance of personal possessions should be kept to a minimum
C. proper, chiefly because the chance of damage is reduced
D. improper, chiefly because the fireman should devote his efforts to putting out the fire

29. The sum of two odd numbers will always be a/an_____ number.

A. composite B. prime C. even D. odd

30. Regulations of fire departments require that when placing hose on a fire apparatus, care should be taken to avoid folding the fire-hose at places where it has been folded already. The most important reason for this requirement is that repeated folding of the hose in the same places will cause:

A. kinks in the hose at those places
B. weakening of the hose at those places
C. discoloration of the hose at those places
D. dirt to accumulate and clog the hose at those places

31. Plain concrete, without reinforcing rods can be used only where the imposed load is:

A. eccentric B. static C. tensile D. compressive

32. Holding a fire drill during the performance in a Broadway theater would be ineffective mainly because:

A. the audience is in the theater for pleasure and would resent interruption of the performance
B. the members of the audience are transitory and do not return to this theater regularly
C. the members of the audience have easy access to many exits in the theater permitting speedy and safe evacuation of the building
D. theaters are usually of fireproof construction and unlikely to catch fire

33. Suppose you are a firefighter making an inspection of a business. During the inspection, the business owner asks you a technical question which you cannot answer. Of the following, the best procedure for you to follow is to:

A. tell him you are not there to answer his questions but to make an inspection
B. guess at the answer so that he won't doubt your competency
C. tell him you don't know the answer but you will look it up and notify him
D. give him the title of a textbook that probably would contain the information

34. Of the following, which is the greatest cause for the loss of firefighters lives, the lack of:

A. proper drills
B. equipment
C. manpower
D. knowledge of structures

35. The exit doors in a theater should swing out in the direction of the street mainly because:

A. panic should be avoided
B. people should walk, not run, to the nearest exit
C. the doors may catch fire
D. audiences can then get out easier

36. A manager of a large apartment building discovers a fire in a vacant apartment. After notifying the fire department, he goes to the basement and shuts off the central air-conditioning system. In so doing, the superintendent acted:

A. wisely; escape of gas fumes from the air-conditioning system is prevented
B. unwisely; the fire would have been slowed down by the cooling effect of the air conditioning
C. unwisely; the air-conditioning system would have expelled smoke from the building
D. wisely; spread of the fire by means of a forced draft will be prevented

37. By increasing the area of contact, wetting agents added to water:

A. Increase the surface tension
B. increase the rate of heat transfer
C. decrease the rate of heat transfer
D. decrease the penetrating absorption

38. The term used to express electrical pressure is:

A. volt B. ampere C. pressure gauge D. ohm

39. A fire will occur when a flammable material is sufficiently heated in the presence of oxygen. It could be assumed from the previous statement that a fire will occur whenever there is sufficient:

A. heat, hydrogen, and fuel
B. oxygen, hydrogen, and fuel
C. heat, air, and fuel
D. gasoline, heat, and fuel

40. Which of the following is <u>not</u> represented by a side of the fire triangle:

A. fuel
B. uninhibited chemical chain reaction
C. temperature
D. oxygen

41. Which one of the following materials is considered to be the most fire resistant:

A. rubber
B. fiber board
C. cork
D. asbestos
E. cellulose acetate

42. In general, solid materials do not transmit or retransmit heat readily by:

A. radiation
B. conduction
C. convection
D. reflection
E. absorption

43. Generally speaking, before a fire can be initiated there always must be:

A. nitrogen
B. organic materials
C. oxygen
D. a catalyst
E. radiation

THE QUESTIONS BELOW REFER TO THE FOLLOWING DRAWING. WHEELS "A" AND "B" HAVE THE SAME DIAMETER. WHEEL "C" HAS A DIAMETER 1/3 THE DIAMETER OF WHEEL "A". WHEEL "A" IS CONNECTED TO WHEEL "C" BY A BELT AND WHEEL "B" IS FIXED TO WHEEL "C". WHEEL "A" ROTATES IN A CLOCKWISE DIRECTION AT A SPEED OF 10 RPM. (questions #44 & #45)

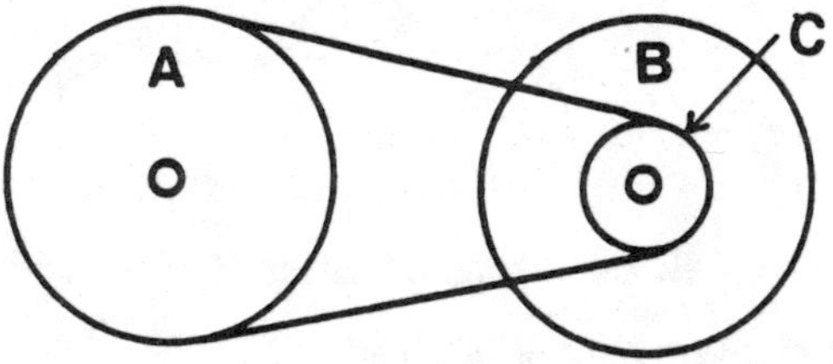

44. If wheel A rotates in a clockwise direction then:

A. wheels C and A turn in opposite directions
B. Wheels B and A turn the same direction
C. wheels C and B turn in opposite directions
D. wheels A and B turn clockwise and C turns counterclockwise.

45. If wheel A turns at a rate of 10 rpm, the number of rpm made by wheel C is most nearly:

A. 90 rpm
B. 60 rpm
C. 30 rpm
D. 3 rpm

46. In the drawing below, while lifting the load the required pull decreases:

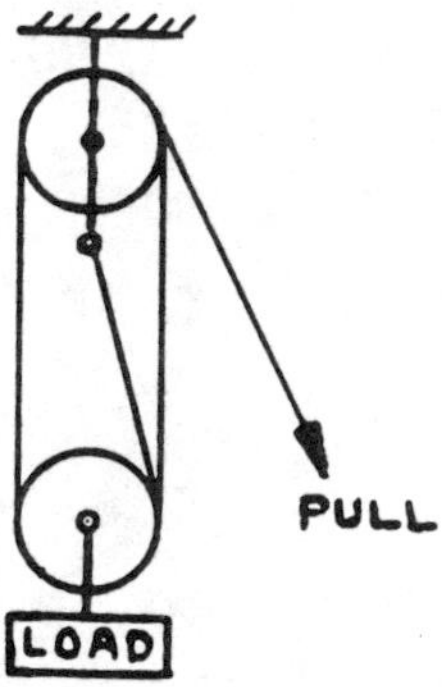

A. as the distance between the pulley wheels decreases
B. when the rope size is increased
C. when the number of pulleys is increased
D. if larger pulley wheels are used

47. If gear X is rotated in the direction of the arrow, which direction will gear Y rotate?

A. clockwise B. counterclockwise C. both directions

48. The object shown below is used to:

A. scrape paint
B. etch leather
C. install window glass
D. drives screws in close areas

49. The object shown below is used to:

A. start engines
B. drill holes
C. replace spark plugs
D. change auto tires

50. The object shown below is a:

A. paper clip
B. wire staple
C. cotter pin.
D. finishing pin

WHICH NUMBER LOGICALLY FOLLOWS IN EACH NUMERICAL PROGRESSION: (questions #51 - #54)

51. 7 10 8 12 9 12 10 ___

A. 12 B. 14 C. 16 D. 18

52. 4 4 0 5 5 1 6 ___

A. 4 B. 5 C. 6 D. 8

53. 67 59 64 16 48 40 ___

A. 45 B. 65 C. 48 D. 18

54. 5 6 7 8 16 17 34 ___

A. 9 B.18 C. 35 D. 68

WHICH LETTER COMES NEXT IN THESE ALPHABETICAL PROGRESSIONS: (questions #55 - #60)

55. A D G J ___

A. K B. L C. M D. O

56. G H I J K _

A. L B. M C. N D. O

57. A C B D C E _

A. C B. D C. B D. E

58. B E J Q _

A. Z B. T C. W D. A

59. H J L N _

A. M B. O C. P D. Q

60. G H I J K _

A. L B. M C. N D. O

61. JUSTIFY means most nearly:

A. shield B. appreciate C. complete D. defend

62. To free from blame is to:

A. implicate B. exculpate C. resign D. tolerate

63. Wire connecting the two terminals must be kept taut. As used in this sentence, the word taut means most nearly, without:

A. defects B. slack C. pressure D. electrical charge

64. A man who builds with stone, brick, or similar materials is called a(n):

A. artisan B. mason C. laborer D. journeyman

65. To swing backward and forward defines the word:

A. alternate B. gesticulate C. oscillate D. procrastinate

STATEMENT

Within the fire service, the Hurst tool and shears have many uses in both rescue and forcible entry. Each tool, the "jaws of life" and the shears, may be used individually or simultaneously from the same power supply. The power supply produces hydraulic oil pressure that circulates through flexible lines. These lines must be attached to the tools before they will function. The fluid used for the tools is not a standard hydraulic oil, it is an oil of an acid base that can cause burns to the skin if contacted. Each tool weighs about 60 lbs. The power unit weighs about 40 lbs. The flexible arms of the tool are about 2 1/2 feet long and made of forged titanium. The tensile strength of the arms is 155,000lbs. The actual force exerted at the tip of the "jaws" is 10.000 lbs in both the push or pull operation. The "jaws of life" are used for prying and separating, without the ability to perform cutting operations. To operate the tool firefighters must grasp the back portion with both hands, using the thumb on the right hand to control the opening and closing of the jaws. The control must also be operated by left handed firefighters with their left thumb.

Firefighters may cut the post, which anchors a car top to the body of the car with one cutting motion with the use of the shears. When this is completed, firefighters may extricate victims from the car by literally peeling the car top back. Firefighters should always be aware of sharp and jagged edges while operating the shears. Using these tools, firefighters can accomplish in seconds cutting operations which include windshield post and metal-clad door frames. Firefighters, with the added use of chains, can pull a steering wheel and column completely out of a car. The tool is capable of lifting the weight of a car or a truck.

ACCORDING TO THE PREVIOUS STATEMENT
(question #66)

66. The hurst tool and shears:

    A. must be operated separately
    B. must be operated together
    C. may be operated separately or together from one power supply
    D. if operated together must have two power supplies

STATEMENT

A flameproof fabric is defined as one which, when exposed to small sources of ignition, such as sparks or smoldering cigarettes, does not burn beyond the vicinity of the source of ignition. Cotton fabrics are the materials commonly used that are considered most hazardous. Other materials such as acetate rayons and linens are somewhat less hazardous, and woolens and some natural silk fabrics, even when untreated, are about equal to the average treated cotton fabric in so far as flame spread and ease of ignition are concerned. The method of application is to immerse the fabric in a flame-proofing solution. The container used must be large enough so that all the fabric is thoroughly wet and there are no folds which the solution does not penetrate.

ACCORDING TO THE PREVIOUS STATEMENT
(questions #67 - #69)

67. A flameproof fabric is one which:

A. is unaffected by heat and smoke
B. resists the spread of flames when ignited
C. burns with a cold flame
D. cannot be ignited by sparks or cigarettes
E. may smolder but cannot burn

68. Woolen fabrics which have not been treated are as likely to catch fire as:

A. treated silk fabrics
B. treated acetate rayon fabrics
C. untreated linen fabrics
D. untreated synthetic fabrics
E. treated cotton fabrics

69. In the method described above, the flame-proofing solution is best applied to the fabric by:

A. sponging the fabric
B. spraying the fabric
C. dipping the fabric
D. brushing the fabric
E. sprinkling the fabric

STATEMENT

During search operations the first step is usually to rescue victims who can be seen and heard, or those whose exact locations are known. Disorganized or careless search must be avoided since victims may be underneath rubble. Disorganized movement could cause injury or death. The best method is to start from the outer edge and work toward the center of an area. Sometimes a trapped or buried victim may be located by calling out or by tapping on pipes. Rescue workers should first call out, then have a period of silence to listen for sounds form a victim.

ACCORDING TO THE PREVIOUS STATEMENT
(questions #70 -#75)

70. The main point of the above statement is:

A. firefighters should call and listen often
B. trapped victims can usually be heard
C. searching should be an organized procedure
D. many victims are buried in fires

71. When searching for buried victims it is very important for firefighters to:

A. have periods of silence
B. keep moving constantly
C. search rubble piles quickly
D. stay away from rubble piles

72. Disorganized movement by a rescue worker can cause:

A. panic and confusion
B. property destruction
C. wasted time
D. death or injury

73. The best way to search an area is:

A. around the edges
B. from center to edge
C. from corner to corner
D. from edge to center

74. Normally the first victims to be rescued during a search are those who are:

A. unconscious
B. trapped under rubble
C. easy to see and hear
D. injured

75. Tapping on pipes is a good way to locate victims because:

A. firefighters can use Morse code
B. sound travels through a pipe
C. firefighters can signal each other this way
D pipes usually are not covered by rubber

STATEMENT

Fire records indicate that about 10% of the total value of fire loss in the United States is due to fires caused by heating equipment, and that insufficient clearance to combustible materials is an important factor in a large percentage of these fires. The need for adequate clearance is emphasized at periods of severe cold weather, when heating equipment is run at full capacity, by a marked increase in the number of fires.

ACCORDING TO THE PREVIOUS STATEMENT
(questions #76 - #78)

76. In a given year in the United States, 2,000,000 fires caused 10,000 deaths and destroyed $500,000,000 worth of property. It can be conclude from the previous statement that heating equipment caused:

A. 200,000 fires
B. fires resulting in the deaths of 1,000 persons
C. fire resulting in the destruction of $500,000,000 worth of property
D. 200,000 fires resulting in the death of 1,000 persons and the destruction of $50,000,000 worth of property

77. The month-to-month changes in the number of fires caused by heating equipment is most affected by the:

A. season of the year
B. amount of clearance to combustible materials
C. condition of the equipment
D. long term trend in fire statistics

78. By "clearance to combustible materials" as used in the above statement the author means:

A. testing laboratory approval for heating equipment
B. maintaining heating equipment in clean condition
C. keeping sufficient space between heating equipment and inflammable material
D. obtaining appropriate permits before installing the heating equipment

79. A situation involving choices between equally unsatisfactory alternatives is called a:

A. crisis B. deadlock C. dilemma D. farce

80. To make rows of small holes through a substance defines the word:

A. penetrate B. perforate C. permeate D. pulsate

CHOOSE THE WORD IN EACH QUESTION THAT IS MISSPELLED
(questions #81 & 82)

81.
A. narrow
B. dispute
C. bargan
D. rapidity

82.
A. adjourne
B. hazardous
C. meddling
D. inquisitive

STATEMENT

Fire often travels inside the partitions of a burning building. Many partitions contain wooden studs that support the partitions. The studs leave a space for the fire to travel along. Flames may spread from the bottom to the upper floors through the partitions. Sparks from a fire in the upper part of a partition may fall and start a fire in the bottom. Some signs that a fire is spreading inside a partition are:

1. Blistering paint.
2. Discolored paint or wallpaper.
3. Partition may feel hot to the touch.

If any of these signs are present the partition must be opened up to look for the fire. Finding cobwebs inside the partition is one sign that fire was not spread through the partition.

ACCORDING TO THE PREVIOUS STATEMENT
(questions #83 - #87)

83. It is possible for fires to spread inside partitions because:

A. fire will burn in any place
B. of the spaces between the studs
C. the material that the studs are made of
D. partitions are painted

84. The evidence of cobwebs inside a partition would indicate that a fire has not spread inside the partition, because:

A. fire will envelop cobwebs
B. the spaces between studs will be blocked by cobwebs
C. cobwebs are fire resistive
D. cobwebs are not constructed inside partitions.

85. The principal message of the statement is:

A. the way partitions are constructed
B. how to avoid the spread of fire
C. The value of cobwebs to firefighters
D. how fire spreads inside partitions

86. A method of determining if fire is spreading within a partition is to check the:

A. age of the plaster
B. color of the smoke
C. temperature of the partition
D. spacing of the studs

87. When a firefighter observes the blistering of paint on a partition, the firefighter should:

A. make an opening in the partition
B. soak the partition with large amounts of water
C. inspect all of the partitions in the building
D. inspect the partitions of other rooms within the structure

88. Which of the following sentences is BEST with respect to grammar, usage, punctuation, and spelling:

A. Many of the cadets' drills were supervised by captains who's interest lay in teaching.
B. Many of the cadet's drills were supervised by captains who's interest lay in teaching.
C. Many of the cadets' drills were supervised by captains whose interest lay in teaching.
D. Many of the cadet's drills were supervised by captains whose interest lay in teaching.
E. Many of the cadet's drills were supervised by captains' whose interest lay in teaching.

89. Which of these sentences is grammatically **INCORRECT?**

A. She wrung her hands in distress.
B. He brought the package home last night only to take it back this morning.
C. He should have lain quietly until the doctor came.
D. My reason for refusing to go is that I have no money.
E. Her brother graduated high school last year.

90. Which of these sentences is grammatically **INCORRECT?**

A. Whoever wins the election will receive our support.
B. I cannot but praise you for your courage.
C. You must either decide to attack at once or to retreat in the morning.
D. The sign read "Boys' and Children's Clothing."
E. The data which were used had been supplied by the agents.

91. Choose the misspelled word

A. idiomatic
B. ingenuity
C. uncanny.
D. none misspelled.

92. Choose the correct spelling

A. satesfaction
B. satisfaction
C. satisfackion
D. none of these

93. Arterial pressure points are:

A. deep seated and require great pressure
B. best located by taking a pulse
C. close to bones near the surface of the body
D. used to cut off all blood circulation

94. A fracture is an injury which occurs mainly in the:

A. nerves
B. muscles
C. bones
D. ligaments

95. An agent used to prevent or counteract the action of a poison is called an:

A. antibiotic B. antibacterial C. antacid D. antidote

96. When a firefighter suffers a burn at a fire, the direct cause of this burn is most likely:

A. flame B. carbon monoxide C. smoke D. heat

97. The foam in foam fire extinguishing systems, being composed largely of water, has a cooling or wetting effect which is effective in the extinguishment of glowing embers. Foam has the property of insulating solids or liquids against re-ignition for a considerable period after its application, and has the advantage that it is not likely to be carried away by the draft of the fire, by natural drafts, or by drafts produced by ventilating systems. According to the previous statement:

A. foam is a volatile substance
B. solid matter treated with foam is not prone to combustion
C. a ventilating system invariably produces a draft
D. foam is an excellent insulating substance

STATEMENT

The automatic sprinkler is a device for automatically distributing water upon a fire in sufficient quantities either to extinguish it entirely or hold it in check in case the fire is located where it is impossible for the water spray to reach it.

ACCORDING TO THE PREVIOUS STATEMENT

98. The automatic sprinkler:

A. is a most effective device in the firefighting program
B. is an excellent fire retardant
C. works like an automaton
D. operates with water as its sole extinguishing agent

99. As a witness to a man falling off the roof of a one story home and becoming unconscious, of the following actions what should you do:

A. hold his head up and give him a stimulant
B. move him inside the house
C. call for help and do not move the man
D. walk him around until he awakens

100 Of the following, which is the correct First Aid procedure for Chemical burns:

A. apply baby oil solution to the burn
B. apply a dry dressing
C. dry dressing after thorough flushing with water
D. flush with water and leave uncovered

PRACTICE EXAM #2
ANSWERS

1. D
2. C
3. D
4. A
5. D
6. D
7. D
8. D
9. B
10. D
11. B
12. B
13. D
14. B
15. B
16. C
17. D
18. A
19. E
20. A
21. C
22. B
23. A
24. B
25. C
26. A
27. A
28. C
29. C
30. B
31. D
32. B
33. C
34. D
35. D
36. D
37. B
38. A
39. C
40. B
41. D
42. C
43. C
44. B
45. C
46. A
47. A
48. D
49. D
50. C
51. B
52. C
53. A
54. C
55. C
56. A
57. B
58. A
59. C
60. A
61. D
62. B
63. B
64. B
65. C
66. C
67. A
68. E
69. C
70. C
71. A
72. D
73. D
74. C
75. B
76. C
77. A
78. C
79. C
80. B
81. C
82. A
83. B
84. A
85. D
86. C
87. A
88. D
89. E
90. C
91. D
92. B
93. C
94. C
95. D
96. D
97. B
98. D
99. C
100. C

PRACTICE EXAM
#3

FIREFIGHTER WRITTEN
PRACTICE EXAM #3

1. Of the choices below, which is the one that describes the reason for insulating steam pipes used in heating systems?

   A. To prevent structure fires.
   B. To prevent corrosion from forming within the pipes.
   C. To prevent the loss of heat.
   D. To prevent the chance of occupants from being burned.

2. In a room filled with smoke, the best air will normally be found near a/the:

   A. window B. doorway C. ceiling D. floor

3. Add:

   53 + 26 + 17 =

   A. 86 B. 89 C. 93 D. 96

4. Subtract:

   77 - 65 =

   A. 12 B. 14 C. 15 D. 16

5. Multiply:

   21 X 44 =

   A. 904 B. 914 C. 924 D. 925

6. Divide:

   42 by 7 =

   A. 4 B. 5 C. 6 D. 8

7. Find the percentage:

   20% of 30 =

   A. 6 B. 15 C. 3 D. 10

8. Firefighter Simpson went to the market with a $20 bill. He bought the following items: roast, $7.87; potatoes, $2.34; milk, $l.88; salad fixings $4.65; ice cream, $2.69. How much change did he bring back to the fire station?

   A. 57 cents B. 61 cents C. 66 cents D. 87 cents

9. During one year the City Fire Departments HQ station had 5,566 total responses. During the same period Station #2 had 6,434 total responses. How many more responses did Station #2 have than HQ?

   A. 686 B. 688 C. 866 D. 868

10. How much longer does it take a ladder truck to travel one mile at 20 miles per hour than at 30 miles per hour?

A. 1 min. B. 10 min. C. 20 min. D. 40 min.

IN THE FOLLOWING DRAWINGS DETERMINE THE RELATIONSHIP BETWEEN THE FIGURES LOCATED IN THE TOP TWO BOXES AT THE LEFT, THEN DETERMINE HOW THE FIGURE IN THE BOTTOM LEFT BOX IS SOMEHOW RELATED TO ONE OF THE FOUR FIGURES TO THE RIGHT: (REMEMBER POSITION & DIRECTION)

11.

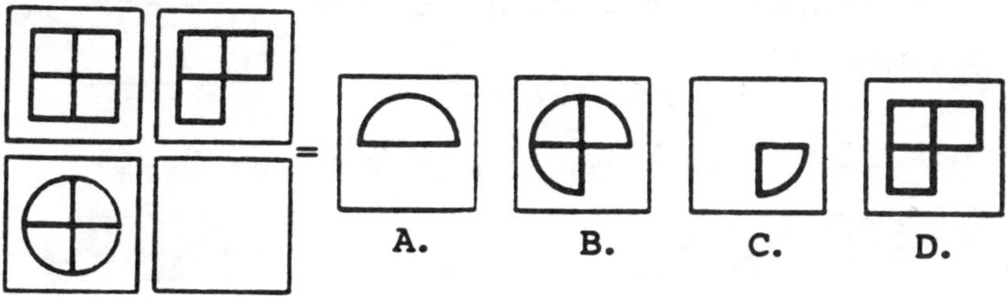

12.

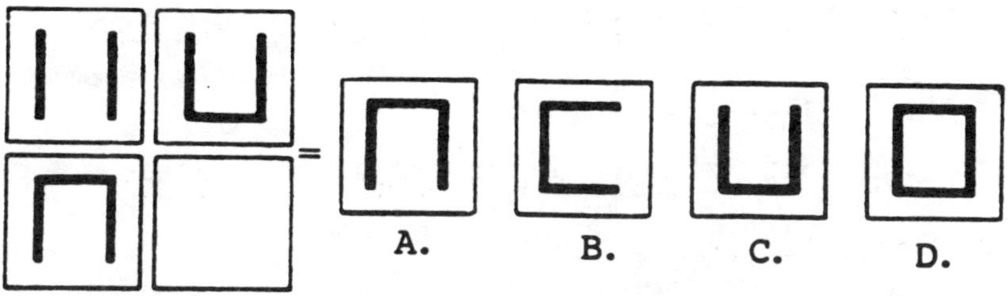

13.

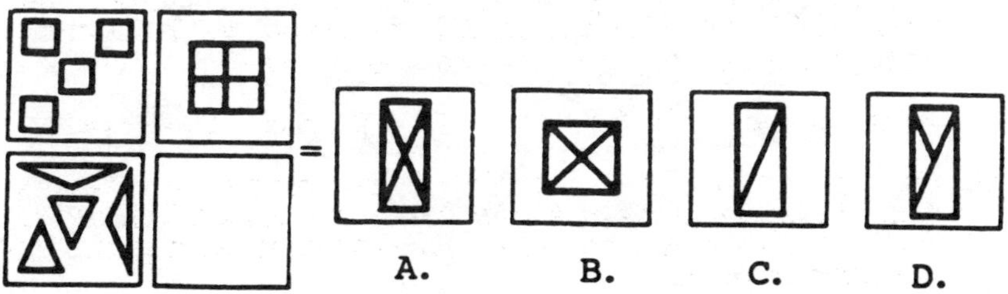

CUBES:

IN THE FOLLOWING QUESTIONS (14 & 15) PERTAINING TO CUBES/BLOCKS, IN EACH PARTICULAR DRAWING THE CUBES/BLOCKS ARE OF THE IDENTICAL DIMENSIONS AND SHAPE; ALTHOUGH THE DIMENSIONS AND SHAPE MAY VARY FROM ONE QUESTION TO ANOTHER. IF NECESSARY, DON'T FORGET TO COUNT UNSEEN BOXES NEEDED TO SUPPORT DRAWINGS:

14. How many blocks are there in the drawing below?

A. 6 Blocks.
B. 8 Blocks.
C. 9 blocks.
D. 10 blocks.

15. How many blocks are there in the drawing below?

A. 10 Blocks.
B. 12 Blocks.
C. 13 blocks.
D. 15 blocks.

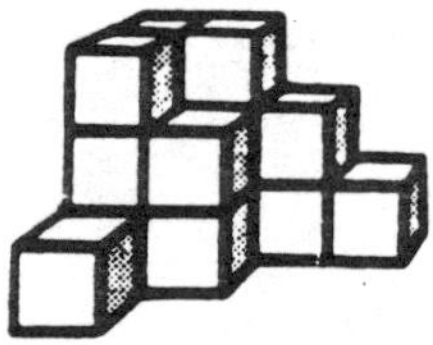

16. Of the choices below, which would be the most probable cause of the tires on the front wheels of a fire rescue apparatus wearing out much faster than those on the rear wheels?

A. Hitting curbs with the front tires when fire rescue apparatus is going around corners.
B. The wheels are improperly aligned.
C. Improper driving of the fire apparatus.
D. The engine of the fire rescue apparatus is located in the front causing the front wheels to carry more weight than the rear wheels.

17. In the drawing below, if the cylinder is rotated in the direction of the arrow, in what direction will the pully hook go?

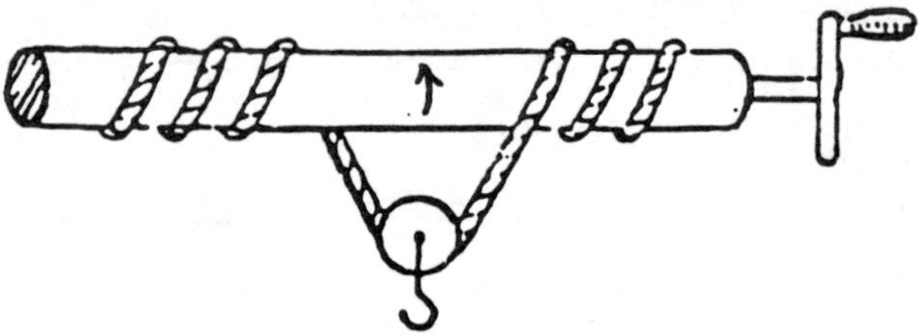

A. Downward   B. Upward   C. neither

18. In which direction must the larger gear rotate in order to make the "C" rotate?

A. Counterclockwise B. Clockwise. C. Both

19. In the drawing below, which teeter would be the most stable?

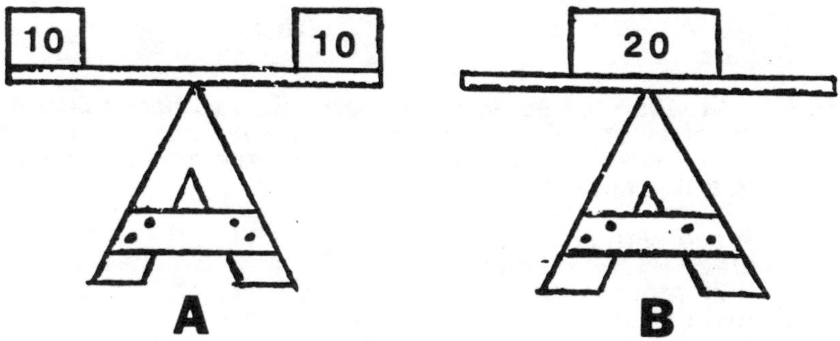

A. Teeter A B. Teeter B C. equally stable

20. The object shown below is a:

A. drawer pull
B. cleat
C. door stop
D. window lock

21. Steel supporting beams in buildings often are surrounded by a thin layer of concrete to keep the beams from becoming hot and collapsing during a fire. The one of the following statements which best explains how collapse is prevented by this arrangement is that concrete:

A. becomes stronger as its temperature is increased
B. acts as an insulating material
C. protects the beam from rust and corrosion
D. reacts chemically with steel at high temperatures

22. The object shown below is used to:

A. measure outside dimensions
B. measure inside dimensions
C. draw circles
D. mark length

23. Firefighters usually lean forward when using a charged hose line. What is the best reason for this?

A. The firefighter is more comfortable because of the cooling of the surrounding air at the tip
B. A backward force is developed which must be counteracted
C. The firefighters are better protected
D. the firefighters can see where the stream strikes better from this position

24. Of the following valves, the one that should be used either entirely open or fully closed, is:

A. globe
B. gate
C. pressure
D. reducing

25. In a situation where an electrician needs to flare the end of a piece of electrical conduit to create a coupling with another piece of electrical conduit, which of the tools listed below should be used?

A. brace with an drill bit
B. brace with a reamer
C. drill motor with a spade bit
D. drill motor with a countersink

26. A priming valve would most likely be used in connection with:

A. standpipe
B. hydrant
C. pump
D. sprinkler system

27. The rock formed largely from shells and skeletons of sea animals is:

A. limestone
B. gneiss
C. sandstone
D. marble

28. The particle having the least weight in the make-up of an atom:

A. a neutron
B. a proton
C. an electron
D. the particle having the least weight in the make-up of an atom has not yet been identified by scientists

29. A parapet wall is the :

A. exterior wall exposed to weather
B. portion of a wall extending above the roof
C. connecting wall between floors
D. interior fire wall between connecting buildings

30. What is the advantage of wired glass over plain glass from a fire protection standpoint:

A. wired glass is more conductive and passes off the heat
B. wired glass stays in position even when cracked
C. heat does not crack wired glass
D. it is more easily replaced

INSERT THE CORRECT LETTER THAT WILL COMPLETE THE FOLLOWING ALPHABETICAL PROGRESSIONS: (questions 31 - 33)

31. A D G J __

A. K B. L C. M D. O

32. B E J Q __

A. W B. X C. Y D. Z

33. A C B D C E __

A. D B. E C. F D. G

IN THE FOLLOWING NUMBER PROGRESSIONS, ADD THE CORRECT NUMBER THAT SHOULD BE PLACED AT THE END OF EACH SERIES: (questions 34 - 39)

34. 9 12 15 18 21 24 __

A. 25 B. 26 C. 27 D. 28

35. 2 6 12 20 30 42 56 __

A. 62 B. 72 C. 82 D. 92

36. 4 4 0 5 5 1 6 __

A. 2 B. 7 C. 6 D. 10

37. 10 30 32 16 48 50 25 __

A. 62 B. 85 C. 75 D. 55

38. 11 2 12 3 13 4 14 __

A. 2 B. 5 C. 7 D. 9

39. 4 8 16 8 16 32 25 __

A. 36 B. 38 C. 40 D. 48

40. Of the following, what treatment should be given to a patient that has a laceration of about two inches long on the right leg that is bleeding moderately:

A. apply pressure directly to the wound
B. apply a tourniquet between the wound and the heart
C. apply pressure to the pressure point directly above the wound
D. allow the wound to bleed so as to avoid infection

41. Of the following choices, which is the best First Aid treatment for a patient having a seizure:

A. give patient water
B. prevent patient from injuring himself
C. hold patient down
D. hold patients mouth open

42. The name which is applied to the attraction that exists between the earth and bodies upon or near it is termed:

A. vibration
B. balance of force
C. leverage
D. gravity

43. On a firefighters first day on the job, a neighborhood resident visits the firehouse and complains to the firefighter that too much of the taxpayers' money is being spent on expensive firefighting equipment. The firefighter should say:

A. "I'm new on this job so let me get my superior officer to discuss this with you."
B. "Since New York City is the biggest city, it needs the most expensive equipment."
C. "Sir, you are entitled to your opinion even if you are wrong."
D. "Why don't you write a letter of complaint and send it to the mayor?"

44. In cases of suspected arson, it is important that firefighters engaged in fighting the fire remember conditions that existed at the time of their arrival. Particular attention should be given to all doors and windows. The main justification for this statement is that knowledge of the condition of the doors and windows may indicate:

A. where the fire started
B. who set the fire
C. the best way to ventilate the building
D. that someone wanted to prevent extinguishment

45. Firefighters frequently open doors, windows, and skylights of a building on fire in a planned or systematic way in order to ventilate the building. The one of the following which is least likely to be accomplished by ventilation is the:

A. increase in visibility of firefighters
B. slowdown in the rate of burning of the building
C. reduction in the danger of toxic gases
D. control of the direction of travel of the fire

46. Of the following the best way to slow down the spread of fire from one room to another is to:

A. close the door between rooms
B. remove contents of rooms
C. close all the windows in the room
D. put holes in the walls between rooms

47. Fires during high humidity conditions tend to produce which of the following:

A. smoky fires
B. fires in which the smoke will rise very fast
C. fires that very easy to fight
D. very intense fires

48. Fires involving flammable gasses are classified as:

A. Class A
B. Class B
C. Class C
D. Class D

49. A man is an uncle when he has a:

A. mother
B. wife
C. brother
D. niece
E. son

50. The fire triangle is a three sided figure that represents the three factors that are necessary for combustion. Which of the following is not one of these factors:

A. oxygen
B. carbon monoxide
C. heat
D. fuel

51. A firefighter's mental attitude is of vital importance because:

A. he must keep calm in order to think clearly in emergencies
B. firefighting often calls for executive ability
C. intelligence as well as physical stamina is needed
D. his efficiency on the job is always visible to the public

52. The best way for a firefighter to find his way out of a room filled with dense smoke is to:

A. crawl along the floor using a flashlight
B. feel his way along the wall until he finds an opening
C. follow the hose line that has been led into the building
D. go in the direction of outside noises

53. Firefighters are required to wear steel reinforced, inner soles inside their rubber boots. The main purpose of these inner soles is to:

A. make the boots more durable and long lasting
B. protect the firefighters feet from burns from smoldering objects or embers
C. protect the firefighters feet from injury from falling objects
D. protect the firefighters feet from nails or other sharp objects

54. While performing a routine inspection of a factory building, a firefighter is asked a question by the plant manager about a matter which is under the control of the city's health department and about which the firefighter has little knowledge. In this situation, the best of the following courses of action for the firefighter to take is to:

A. answer the question to the best of his ability
B. tell the manager that he is not allowed to answer the question because it does not relate to a fire department matter
C. tell the manager that he will refer the question to the health department
D. suggest to the manager that he communicate with the health department about the matter

55. No firefighter is expected to use a measuring stick to determine the exact amount of hose needed to stretch a line from the entrance of a burning building to the seat of the fire. He should be guided generally by he following rule-one length of hose for every story. This rule assumes most directly some degree of uniformity in:

A. fire hazard
B. window areas
C. building construction
D. street widths

56. A shaft is best defined as being:

A. an enclosed space within a building connecting a series of two or more openings in successive floors
B. A space enclosed on all sides with masonry or reinforced concrete and used for transmission of air or smoke
C. a passage of sheet metal used as a connection between a heater and chimney for the transmission of smoke
D. a subdividing partition that is built to provide an area of refuge in a building

57. An intense fire develops in a room in which carbon dioxide cylinders are stored. The principal hazard in this situation is:

A. the CO2 might catch fire
B. toxic fumes might be released
C. the cylinders may explode
D. released CO2 may intensify the fire

58. A floating object will float deeper in fresh water than in salt water because:

A. in the salt water, the salt will occupy some of the space
B. fresh water is heavier than salt water
C. salt water is heavier than fresh water
D. salt water offers less resistance than fresh water

59. When coal burns in a furnace, the weight of all the substances derived from the burning will be equal to the weight of:

A. the coal
B. the ashes taken from the furnace
C. the air entering the furnace plus the weight if the ashes
D. the oxygen entering the furnace + the weight of the coal

60. On various occasions firefighters are required to break windows so that they may gain entry into a structure. With safety being a primary factor, which of the following procedures would **NOT** be recommended?

A. taking a position on the leeward side of the window
B. positioning hands above the point of impact
C. the firefighter not using hans to break the glass
D. start at the top of the window and work way to the bottom

61. Assume that you have two containers of equal size and shape. They are filled with equal amounts of water. A block of ice is added to one container and the same quantity of ice chopped into cubes is added to the other container. What will happen to the water in the container with the cubes as compared to the container with the block:

A. it will cool to the same temperature more quickly
B. it will cool to the same temperature more slowly
C. it will cool to a lower temperature more quickly
D. it will cool to a lower temperature more slowly

62. Which of the following represents the weight of a gallon of water?

A. 12.5 lbs
B. 8.33 lbs
C. 5.5 lbs
D. 4.33 lbs

63. The occurrence of unusual percentages of carbon monoxide at smoldering fires would most likely be due to:

A. oxygen deficiency
B. very low temperature
C. very high temperature
D. similarity of materials

64. Which of the following statements about combustion is incorrect?

A. it occurs when a substance combines with oxygen
B. it is accompanied by light and heat
C. it is the same as burning
D. it is commonly known as an explosion

65. An increase of temperature from 10 degrees C to 30 degrees C is equivalent to an increase of how many degrees F:

A. 11 1/9
B. 36
C. 43 1/9
D. 68

66. The ignition coil in an automobile:

A. regulates the flow of current to the lights
B. increases the voltage of the current to the spark plug
C. controls the charging rate of the battery
D. boosts the voltage between the battery and the starting motor

67. Wet wood will usually burn but does not make good timber because:

A. water does not burn
B. so much heat is needed to evaporate the water
C. wet wood has a higher kindling temperature than dry
D. wet wood has a lower kindling temperature than dry wood

68. Of the following items commonly found in a household, the one that uses the most electric current is a/an:

A. a 150-watt light bulb
B. a toaster
C. door buzzer
D. 8 inch electric fan

69. Which one of the following is least likely to explode:

A. a can full of gasoline
B. a can from which gasoline has just been emptied
C. a can half full of gasoline
D. a closed can of water over a hot fire

70. Which one of the following is least associated with firefighting:

A. chemistry
B. hydraulics
C. mechanics
D. geology

71. Hot gasses cause a room to burn by:

A. creating a draft
B. exploding
C. giving off sparks
D. raising the room temperature

WHICH OF THE FOLLOWING SENTENCES UTILIZES PROPER SPELLING:(questions #72 & #73)

72.

A. Sophmores taking behavorial psychology must pay a labratory fee
B. Sophmores taking behavorial psychology must pay a laboratory fee
C. Sophomores taking behavioral psychology must pay a laboratory fee
D. Sophomores taking behavioral psychology must pay a labratory fee
E. Sophomores in taking behavioral psychology must pay a loboratory fee

73.

A. Atheletic heroes often find their studies an unnecessary hinderance
B. Athletic heros often find their studies an unnecessary hinderance
C. Athletic heros often find their studies an unnecessary hindrance
D. Athletic heroes often find their studies an unnecessary hindrance
E. Athletic heroes often find their studies an unneccessary hindrance

EACH OF THE FOLLOWING QUESTIONS CONSISTS OF FIVE SENTENCES. ONE OF THEM CONTAINS AN ERROR IN GRAMMAR, USAGE, OR PUNCTUATION. SELECT THE INCORRECT SENTENCE. (questions 74 & 75)

74.

A. Tom is one of those boys who are inclined to act without thinking.
B. The number of accidents caused by drunken driving is increasing daily.
C. I had rather leave immediately than stay and be insulted again.
D. I am in New York for more than five years and have never visited the Bronx Zoo.
E. Let me out at once.

75.

A. The committee took a definite stand which amounted to an objection to the ladies' using the facilities on week ends.
B. If she, who is a teacher for twelve years, has difficulty with the child, how can you, a mere beginner, hope to handle him successfully?
C. If you hadn't lain there so long after you were warned to get up, you would not have caught cold.
D. Will you be ready by the time I arrive?
E. She laid her head on his shoulders.

FIND THE MISSPELLED WORD IN THE FOLLOWING GROUPS OF WORDS: (questions #76 -#82)

76.

A. deliscious
B. enormous
C. likeness
D. witnesses
E. none misspelled

77.

A. sophmore
B. duly
C. lovable
D. across

78.

A. inferred
B. whisle
C. jovial
D. conscript
E. none misspelled

79.

A. amplify
B. greace
C. cholera
D. perilous
E. none misspelled

80.

A. cavalier
B. transparent
C. perjury
D. vivinaty
E. none misspelled

81.

A. cannon
B. hospital
C. distilliry
D. righteous
E. none misspelled

82.

A. mathametics
B. automatic
C. atmospheric
D. sprinkler
E. none misspelled

CHOOSE THE CORRECT SPELLING OF THE FOLLOWING WORDS:
(questions #83 & #84)

83.
A. maxamum
B. maximum
C. maxumum
D. none of the above

84.
A. mercontile
B. mercantile
C. mercuntile
D. none of the above

85. To reduce swelling, apply:

A. hot applications
B. cold applications
C. an electric heating pad
D. a snug bandage

86. Suppose you are called on to give first aid to several victims of an accident. First attention should be given to the one who is:

A. bleeding severely
B. groaning loudly
C. unconscious
D. vomiting

87. One of the commonest faults in the operation of the average fire department is the failure to employ large streams where big fires are being fought. Small streams have neither the range nor the cooling effect necessary in handling big, hot fires. The previous statement best illustrates the fact that:

A. lack of adequate water pressure results in disastrous fires
B. big fires are best controlled by the invariable use of water
C. size of fire streams should generally vary directly with the size of the fires
D. it is wisest policy to use a larger stream than a smaller stream at the majority of fires

88. The same care should be taken to avoid damage when extinguishing a fire in the home of a poor family as would be taken in the home of a wealthy family. The best justification for this statement is that the possessions of a poor family:

A. probably are not covered by fire insurance
B. sometimes include items of great monetary value
C. may be very precious to the family
D. may contain valuable or irreplaceable documents such as birth certificates, military discharge certificates, etc.

89. The path followed by one heavenly body in its revolution about another is called a(n):

A. comet B. meteor C. orbit D. satellite

90. Forthright means most nearly:

A. always right
B. rapidly
C. unusual
D. outspoken
E. overcome

91. Which word does not belong in the following group:

A. hose
B. line
C. coupling
D. nozzle
E. ladder

92. To say that an act has been performed legally means that:

A. it complies with the law
B. it has been ordered by a judge
C. it is not generally known
D. it is performed in a court room
E. it should be reported to the police

93. In changing into vapor, gasoline expands so enormously that one gallon will produce many hundreds of cubic feet of gas. It can best be inferred from the preceding statement that:

A. gasoline is an extremely expansive liquid
B. as gasoline vapor condenses into a liquid, its volume is appreciably decreased
C. a violent chemical reaction occurs upon ignition of a large volume of gasoline vapor
D. the vapor content of gasoline increases inversely with it's expansion

94. Confuse means most nearly:

A. tire
B. bewilder
C. criticize
D. forget

95. Sufficient means most nearly:

A. counterfeit
B. enough
C. scanty
D. too little
E. too much

96. Penetrate means most nearly:

A. enter into
B. dent
C. rebound
D. pass over
E. deflate

97. Impartial means most nearly:

A. loyal
B. capable
C. kind
D. good-humored
E. fair

98. Straightforward means most nearly:

A. hasty
B. frank
C. conflicting
D. helpful
E. solemn

99. Ordinarily means most nearly:

A. seldom
B. without fail
C. necessarily
D. hardly ever
E. usually

100. Abruptly means most nearly:

A. suddenly
B. soon
C. deliberate
D. mysteriously
E. absent-mindedly

PRACTICE EXAM #3
ANSWERS

1. C
2. D
3. D
4. A
5. C
6. C
7. A
8. A
9. D
10. A
11. B
12. D
13. A
14. C
15. D
16. B
17. C
18. C
19. B
20. D
21. B
22. B
23. B
24. B
25. B
26. C
27. A
28. C
29. B
30. B
31. C
32. D
33. A
34. C
35. B
36. C
37. B
38. B
39. D
40. A
41. B
42. D
43. A
44. D
45. B
46. A
47. A
48. B
49. D
50. B
51. A
52. C
53. D
54. D
55. C
56. A
57. C
58. C
59. D
60. A
61. A
62. B
63. A
64. D
65. B
66. B
67. C
68. B
69. A
70. D
71. D
72. C
73. D
74. D
75. B
76. A
77. A
78. B
79. B
80. D
81. C
82. A
83. B
84. B
85. B
86. A
87. C
88. C
89. C
90. D
91. E
92. A
93. B
94. B
95. B
96. A
97. E
98. B
99. E
100. A

PRACTICE EXAM #4

FIREFIGHTER WRITTEN

PRACTICE EXAM #4

1. Add:

   28 + 14 + 37 =

   A. 77 B. 79 C. 87 D. 88

2. Add:

   68 + 23 + 95 =

   A. 176 B. 181 C. 186 D. 286

3. Multiply:

   67 X 8

   A. 528 B. 536 C. 538 D. 543

4. Subtract:

   907 - 214 =

   A. 683 B. 684 C. 783 D. 693

5. Add:

   926
   388
   683

   A. 988 B. 1897 C. 1986 D. 1997

6. A room has an area of 168 square feet. If it's length is 14 feet what width salvage cover will be needed to protect the floor?

   A. 6 feet wide
   B. 8 feet wide
   C. 10 feet wide
   D. 12 feet wide

7. During his first eight-hour day, fire cadet Smith earned 40% as much as firefighter Jones. If the firefighter earned \$28.00, what was the cadet's average hourly earning?

   A. \$1.60 B. \$2.40 C. \$1.68 D. none of these

8. Assume that one-half of a fire department is composed of firefighters, one-third of paramedics, and the rest of apparatus drivers and captains. There are five times as many apparatus drivers as captains. If there are 125 apparatus drivers how many people are there in the entire department?

   A. 500 B. 600 C. 700 D. 900

9. Rewrite 3,756,548 rounded off to the nearest thousand:

A. 3,756,600
B. 3,756,500
C. 3,7756,000
D. none of these

10. Engineer Harper has four lengths of 2 1/2 " hose in the following lengths: 49 feet, 50 feet, 48 feet, and 47 feet. What is the total length of these pieces?

A. 147 feet B. 149 feet C. 194 feet D. 197 feet

11. A stack of "C" collars is 2.75 feet high. If the "C" collars are .375 inch thick, how many are there in the stack:

A. 7 B. 8 C. 14 D. 88

12. If the fuel consumption of a 110 horsepower engine is .75 lb. per horsepower hour, how many pounds of fuel will be used in 40 minutes:

A. .50 B. 30 C. 50 D. 55

13. Multiply:

9.8 X 11 =

A. 107.8 B. 10.78 C. 1.078 D. 1,078

14. Find percentage:

51% of $185. 25 (express to nearest cent)

A. $94.48 B. $279.73 C. $90.77 D. $94.47

15. The more common combustible materials contain:

A. carbon
B. gases
C.oxygen
D. carbon monoxide

16. Ventilation is important in firefighting because it:

A. releases trapped smoke and gases
B. put out flames by cooling them
C. makes it easier for firefighters to breathe
D. makes the flames easier to see and reach with a hose

17. Hot gases are most dangerous when they are in a room that is:

A. large B. closed C. damp D. cool

18. Which one of the following is not one of the ways that heat can travel through a burning building:

A. conduction
B. convection
C. radiation
D. conclusion

19. Water is a better conductor of electricity when:

A. it contains a large amount of organic matter
B. it is absolutely pure
C. its mineral content is high
D. its mineral content is low

20. The chief purpose of insulating steam pipes in heating systems is to:

A. prevent the loss of heat
B. protect people from being burned
C. prevent fires
D. prevent the pipes from rusting

21. The freezing point on a Fahrenheit thermometer is:

A. zero degrees
B. 10 degrees
C. 28 degrees
D. 32 degrees

22. Of the following, what happens to water when it freezes:

A. it contracts
B. it changes chemically
C. it expands
D. it increases substantially

23. A firefighter goes home and finds in his house a very strong and unmistakable odor of gas. Which of the following things should he **NOT** do?

A. leave the house at once
B. be sure that everyone is out of the house
C. before leaving telephone the fire department from the house
D. leave the door open

24. The planet nearest to the earth is:

A. Mercury
B. Pluto
C. Venus
D. Jupiter
E. Saturn

25. Food is preserved by refrigeration because cold:

A. helps to keep food crisp
B. helps to retain moisture
C. retards the growth of bacteria
D. sterilizes the food
E. improves the taste of food

26. How many blocks are there in the drawing below?

A. 8
B. 10
C. 12
D. 16

27. How many blocks are there in the drawing below?

A. 5
B. 6
C. 7
D. 8

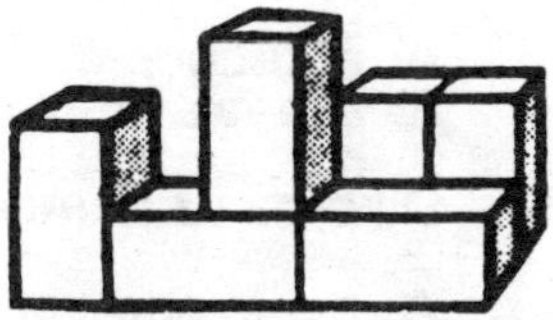

IN THE FOLLOWING DRAWINGS, DETERMINE THE RELATIONSHIP BETWEEN THE FIGURES LOCATED IN THE TOP TWO BOXES AT THE LEFT, THEN DETERMINE HOW THE FIGURE IN THE BOTTOM LEFT BOX IS SOMEHOW RELATED TO ONE OF THE FOUR FIGURES TO THE RIGHT. REMEMBER POSITION AND DIRECTION.

(questions #28 - #30)

28.

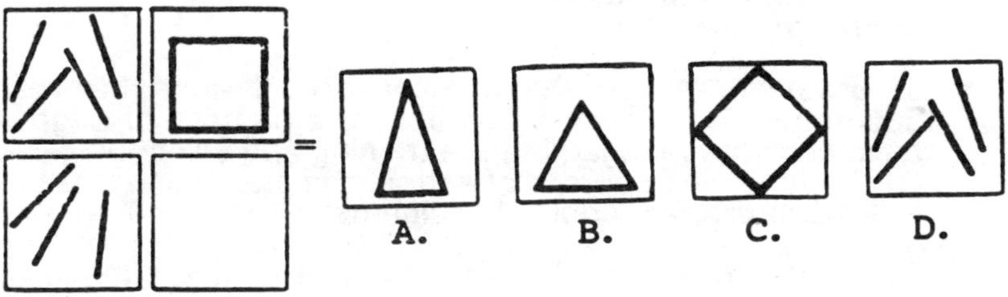

29.

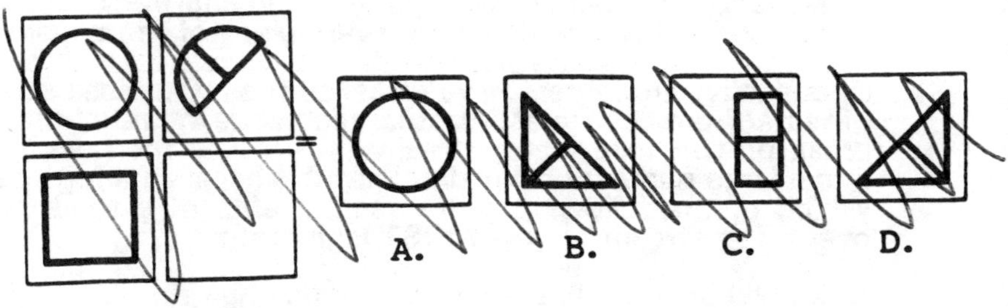

30.

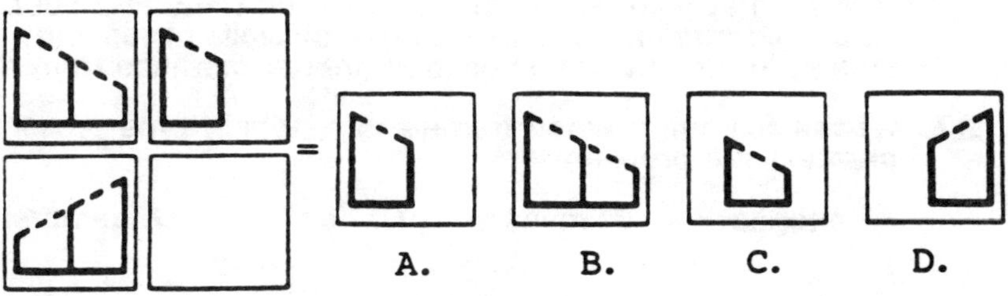

31. What uniform is normally worn while conducting fire prevention inspections?

A. dress uniform
B. work uniform
C. protective clothing
D. work uniform with department insignia

32. A fire prevention officer would be least likely to inspect:

A. churches
B. schools
C. apartment building
D. department stores

33. Disorganized movement by a rescue worker can cause

A. panic and confusion
B. property destruction
C. wasted time
D. death or injury

34. Fire strategy is:

A. the attack on a fire
B. the evaluation of a fire scene
C. the plan of attack on a fire
D. the critique of a fire

35. In order to reduce the number of collisions between fire apparatus and private automobiles, a number of suggestions have been made regarding the training and selection of drivers. Of the following courses of action, the one which is least likely to lead to the reduction of accidents is to:

A. select drivers who have the longest driving experience
B. require all members of a company to drive fire apparatus
C. select drivers who have experience driving trucks
D. select an alternate driver for every piece of fire apparatus

36. Historically, firefighters have attacked fires with solid streams of water from hose lines. Another technique in firefighting is to break up the solid water stream as it leaves the hose nozzle into a large number of tiny droplets; this is called a fog stream. Which of the following professed benefits of a solid stream over a fog stream, is the **MOST** legitimate:

A. a solid stream causes less water damage
B. a solid stream results in less drain on the water supply
C. a solid stream involves less risk of walls collapsing
D. a solid stream can be used at greater distance from the fire

37. Of the following wood frame roofs, which type presents the greatest fire problem:

A. pitched B. mansard C. flat D. gambrel

38. The fie extinguisher rating system consists of a letter designation <u>only</u> for extinguishers intended for use on which type of fire:

1. Class A 2. Class B 3. Class C 4. Class D

A. 3 only
B. 2, 3 & 4
C. 1 & 2
D. 3 & 4

39. The basic assumption of fire prevention educational programs is that people frequently:

A. must be forced into obeying fire laws
B. are unaware of the dangers involved in some of their actions
C. don't care whether or not their actions are dangerous
D. assume that fire insurance protects them against all fire loss

40. The two planets that are referred to as neighbors of earth are:

A. Mars and Jupiter
B. Mars and Venus
C. Venus and Saturn
D. Venus and Jupiter

41. A fireman inspecting a small retail store for hazardous fire conditions is told by the owner that the whole inspection procedure is a waste of time and money. Of the following, the best action for the fireman to take is to:

A. explain to the owner the benefits of the inspection program
B. curtly tell the owner that he is entitled to his opinions and continue the inspection
C. ask the owner if he can suggest a better way of preventing fires
D. continue the inspection without answering the owner

42. A certain fire department recently purchased pumpers equipped with enclosed cabs. In the past, the fire department used apparatus of the open type, with no cab or roof. The main advantage of the enclosed cab is that it provides:

A. additional storage space for equipment
B. a place of shelter for firefighters operating in an area of radioactivity
C. protection for firefighters from weather and injury
D. emergency first aid and ambulance facilities

43. Of the following, which choice is the best reason for smoldering fires to have high percentages of carbon monoxide:

A. the high temperature
B. the low temperature
C. the high amount of oxygen
D. the low amount of oxygen

44. What does the fire tetrahedron represent?

A. the three factors required for combustion
B. the three factors required for fire extinguishment
C. the four factors required for fire extinguishment
D. the four elements required by fire

45. The chief reason an automatic sprinkler system sometimes affords better fire protection than can be secured by a watchman is that:

A. after automatic sprinklers are installed they are always at hand and ready to act
B. automatic sprinklers turn on at periodic intervals
C. even if he saw a small fire start, the watchman could not put it out as quickly as an automatic sprinkler
D. such sprinklers are turned on automatically when the temperature of the room changes even slightly

46. In the drawing below, if pulley "A" rotates in the direction of the arrow, then pulley "B" will rotate:

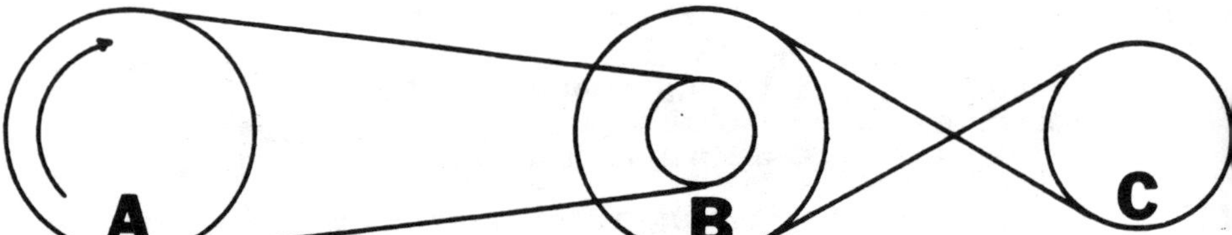

A. slower than A in the opposite direction
B. faster than A in the opposite direction
C. slower than A in the same direction
D. faster than A in the same direction

47. In the drawing below, the small gear will:

A. decrease the speed of gears A and B
B. increase the speed of gears A and B
C. make gears A and B rotate in opposite directions
D. make gears A and B rotate in the same direction

48. Of the following two drawings, which one has the weights that will move?

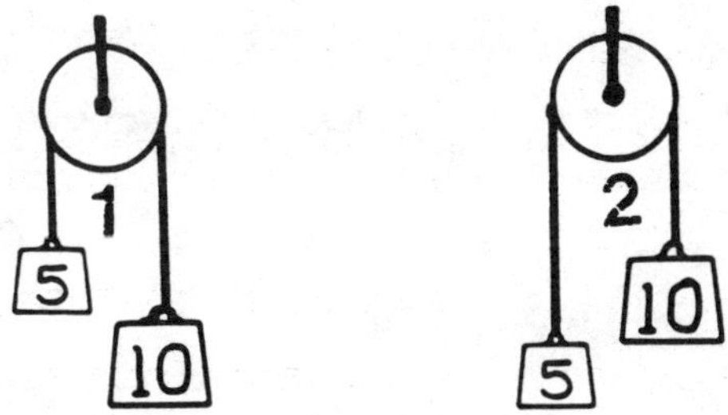

GROUND LEVEL

A. drawing #1
B. drawing #2
C. both drawing #1 & #2
D. neither drawing

49. The object shown below is a:

A. T-square
B. depth gauge
C. combination square
D. guide for a saw

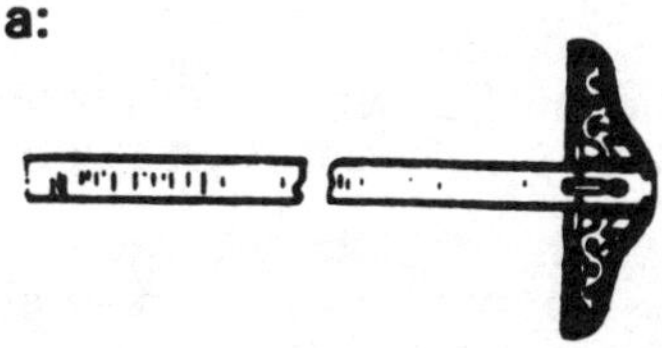

50. The object shown below is used to:

A. thread pipes
B. flare metal tubing
C. pull gears off shafts
D. measure diameters

51. The object shown below is used to:

A. set spark plugs
B. measure saw blades
C. measure drill bits
D. measure wire

52. If the temperature gage in a vehicle indicates that the engine is starting to overheat:

A. Put cold water in the radiator immediately
B. Put hot water in the radiator immediately
C. put antifreeze in the radiator immediately.
D. Allow the vehicle to cool down.

53. The best reason for lubricating moving parts of machines is to:

A. prevent the formation of rust
B. reduce friction
C. increase inertia
D. reduce the accumulation of dirt on parts

54. The type of clutch which gives the most powerful and positive drive is a ________ clutch.

A. jaw B. ring C. disk D. cone

55. Glazing is a term most common to:

A. water color
B. ceramics
C. wood carving
D. etching

56. Gasoline should not be used to clean fire hoses because it would:

A. stain them
B. weaken the couplings
C. shrink them
D. cause them to rot

57. The use of power driven tools can be dangerous. Which of the following would not be a safety technique you should observe?

A. when possible, clamp down the object being worked on
B. use only sharp, clean, well cared for tools
C. approach the operator of the tool only from behind
D. plan how you will proceed with the work before you turn the tool on

58. The term "Phillips head" would be used most appropriately in reference to:

A. a wrench
B. a screwdriver
C. a clamp
D. a saw

59. A wall composed of two different masonry materials so bonded together that they act as a unit under load is a/an:

A. cavity wall
B. enclosure wall
C. faced wall
D. veneered wall

60. The hazard of plywood and similar laminated materials as a contributing factor in flame spread is largely influenced by the:

A. character of the adhesive used between the lamination
B. type of paint used on the outside surface
C. thickness of each laminated layer
D. none of the above

61. Which of the following is considered to be the best conductor of heat:

A. aluminum
B. oak wood
C. brass
D. iron
E. glass

62. Of the following, it would be most accurate to say that when there is smoke there is:

A. an excess of nitrogen
B. an excess of oxygen
C. complete combustion
D. incomplete combustion

WHICH NUMBER COMES **NEXT** IN THE FOLLOWING NUMERICAL PROGRESSIONS? (questions 63 - 67)

63.
3 8 16 21 63 68 136 ___

A. 133 B. 141 C. 146 D. 272

64.
9 10 8 24 6 7 5 ___

A. 10 B. 15 C. 28 D. 32

65.
4 4 0 5 5 1 6 ___

A. 0 B. 2 C. 6 D. 12

66.
35 42 7 12 3 6 3 ___

A. 4 B. 6 C. 12 D. 24

67.
4 3 6 3 12 7 42 ___

A. 4 B. 6 C. 12 D. 35

FIND THE MISSING LETTER IN THE FOLLOWING ALPHABETICAL PROGRESSIONS (questions 68 - 71)

68.
A C E G ___

A. I B. F C. H D. J

69.
K M O Q ___

A. T B. R C. S D. P

70.
Z X V T ___

A. U B. R C. S D. P

71.
J H F D ___

A. E B. C C. B D. G

72. When fighting a fire with water, if the heating effect of the fire is greater than the cooling effect of the water, the later will be decomposed and pass off as steam.

ACCORDING TO THE PREVIOUS STATEMENT

A. steam and water are the same thing
B. fighting a fire with water is ineffective
C. fire may decompose water
D. the heating effect of fire varies inversely with the amount of steam it decomposes

73. WATER is to DELUGE as LAVA is related to:

A. fire
B. eruption
C. Mt. Etna
D. volcano

74. GASOLINE is to AUTOMOBILE as COAL is related to:

A. locomotive
B. burns
C. wheels
D. smoke

75. The plot to set the building afire was thwarted by the department. As used here, the word thwarted means most nearly:

A. blocked
B. discovered
C. punished
D. abetted

76. Religious bigotry is <u>repugnant</u> to all true democrats. As used in this sentence, <u>repugnant</u> means most nearly:

A. dangerous
B. distasteful
C. revealing
D. surprising

77. A sphere of authority is called:

A. constituency
B. dictatorial
C. jurisdiction
D. vassal

78. The officer and his men <u>ascended</u> the stairs as rapidly as they could. As used in this sentence, the word <u>ascended</u> means most nearly:

A. went up
B. washed down
C. chopped
D. shored up

79. <u>Outlet</u> is to <u>electricity</u> as <u>hydrant</u> is to:

A. hose
B. coupling
C. water
D. spanner

80. All stairs and other exits should be readily accessible and so arranged that the path of escape to the street is unmistakable. Signs should be provided to indicate direction of travel to exits where they are not visible from all parts of the floor area.

ACCORDING TO THE PREVIOUS STATEMENT

A. fire prevention is essential
B. fire drill procedures pay off
C. adequate fire precautionary measures are important
D. construction regulations are necessary

81. Some fires are caused by arcing. This term means:

A. an electric current flowing along a conductor bent into a curve
B. an electric current passing across a gap between two conductors
C. the blowing out of an electric fuse
D. a trolley wire coming in contact with the ground rail

82. RAPIDITY means most nearly:

A. power
B. effectiveness
C. possibility
D. swiftness

83. SYNTHETIC means most nearly:

A. artificial
B. hard
C. cheap
D. imported

84. DISPUTE means most nearly:

A. conversation
B. crying spell
C. disagreement
D. performance

85. TOLERANT means most nearly:

A. afraid
B. aware
C. careful
D. understanding

86. ENDURE means most nearly:

A. await
B. bear
C. obey
D. escape

87. Much fire prevention work can be accomplished through inspections by individual fire stations, but the need for specially trained people for a fire prevention unit to conduct inspections of a technical nature is not reduced.

ACCORDING TO THE PREVIOUS STATEMENT

A. an alert mind can take the place of a physical weakness
B. a firefighter should be able to act as well as think
C. inspection by stations will reduce the number of technical inspections
D. technical inspections should be done by specially trained persons

88. Which of the following words is misspelled:

A. pursevere
B. deodorize
C. ligament
D. illuminate
E. none misspelled

89. Which of the following words is properly spelled:

A. pyramaniac
B. spontaneus
C. arsenist
D. vaporize

90. Which of the following words is misspelled:

A. filiment
B. arrestor
C. incinerator
D. flexible

IN THE FOLLOWING QUESTIONS, CHOOSE THE CORRECT SPELLING OF THE WORDS: (questions #91 & #92)

91.

A. difusson
B. diffusson
C. diffusion
D. none of the above

92.

A. directory
B. directery
C. directary
D. none of the above

EACH OF THE FOLLOWING QUESTIONS CONSISTS OF A GROUP OF FIVE SENTENCES. CHOOSE THE SENTENCE THAT IS BEST WITH RESPECT TO GRAMMAR, USAGE, PUNCTUATION, AND SPELLING. (questions #93 - #95)

93.

A. The battle-scarred veteran was as bald as a kneeled egg.
B. The battlescarred veteran was as bald as a new-laid egg.
C. The battle-scarred veteran was as bald as a new-laid egg.
D. The battle scarred veteran was as bald as a new laid egg.
E. The battle-scarred veteran was so bald as a new-laid egg.

94.

A. I get only a week vacation after two years work.
B. I get only a week's vacation after two years work.
C. I get only a week's vacation after two years' work.
D. I get only a weeks vacation after two years work.
E. I get only a week vacation after two years' work.

95.

A. Undoubtedly, even the loneliest patient feels tranquill.
B. Undoubtably, even the loneliest patient feels tranquill.
C. Undoubtedly, even the lonliest patient feels tranquil.
D. Undoubtably, even the loneliest patient feels tranquil.
E. Undoubtedly, even the loneliest patient feels tranquil.

96. Of the following, which medical emergency should be transported the most rapidly:

A. Spine injury
B. Anaphylaxis
C. Stroke patient
D. Heat exhaustion

97. Before entering a room filled with carbon monoxide, firefighters sometimes tie a wet cloth over their mouth and nose. Firefighters should understand that the method used in the previous statement:

A. is ineffective
B. is effective only for impeding asphyxiation
C. will provide protection against carbon dioxide
D. will be effective as long as the handkerchief remains moist

98. Lines of longitude are known as

A. meridians
B. great circles
C. parallels
D. time belts
E. dividers

99. The heart of a normal male adult, at rest, beats approximately:

A. 36 times per minute
B. 70 times per minute
C. 96 times per minute
D. 106 times per minute

100 Of the following methods, which is the best for reducing swelling:

A. cold packs
B. hot packs
C. elastic bandage
D. warm whirlpool

PRACTICE EXAM #4
ANSWERS

1. B
2. C
3. B
4. D
5. D
6. D
7. D
8. D
9. D
10. C
11. D
12. D
13. A
14. A
15. A
16. A
17. B
18. D
19. C
20. A
21. D
22. C
23. C
24. C
25. C
26. C
27. B
28. B
29. A
30. D
31. A
32. C
33. D
34. C
35. B
36. D
37. A
38. D
39. B
40. B
41. A
42. C
43. D
44. D
45. A
46. D
47. BD
48. B
49. A
50. C
51. D
52. D
53. B
54. A
55. B
56. D
57. C
58. B
59. C
60. A
61. A
62. D
63. B
64. B
65. C
66. A
67. D
68. A
69. C
70. B
71. C
72. C
73. B
74. A
75. A
76. B
77. C
78. A
79. C
80. C
81. B
82. D
83. A
84. C
85. D
86. B
87. D
88. A
89. D
90. A
91. C
92. A
93. C
94. C
95. E
96. B
97. A
98. A
99. B
100. A

PRACTICE EXAM

#5

FIREFIGHTER WRITTEN
PRACTICE EXAM #5

EACH OF THE FOLLOWING PROBLEMS CONTAINS A WORD SPELLED SEVERAL WAYS. PICK THE CORRECT SPELLING OF THE WORD. IF NONE OF SPELLINGS ARE CORRECT, CHOOSE THE ANSWER D : NONE OF THESE. (questions 1 - 11)

1.
A. directory
B. directery
C. directary
D. none of these

2.
A. leson
B. lescon
C. lesson
D. none of these

3.
A. collosal
B. collossal
C. colosal
D. none of these

4.
A. actually
B. actualy
C. actuealy
D. none of these

5.
A. relevant
B. relevent
C. rellevent
D. none of these

6.
A. desision
B. desison
C. decision
D. none of these

7.
A. pertinent
B. pertanent
C. pertinant
D. none of these

8.
A. privelige
B. privilege
C. privoledge
D. none of these

9.
A. auxillary
B. auxiliary
C. auxilary
D. none of these

10.
A. horizontal
B. horazontal
C. horizontol
D. none of these

11.
A. hydrent
B. hydrant
C. hydrunt
D. none of these

EACH OF THE FOLLOWING QUESTIONS CONSISTS OF A GROUP OF FIVE SENTENCES. CHOOSE THE SENTENCE THAT IS BEST WITH RESPECT TO GRAMMAR, USAGE, PUNCTUATION, AND SPELLING. (12 - 15)

12.
A. The only man who I think will knock out whoever he fights is Tyson.
B. The only man who I think will knock out whomever he fights is Tyson.
C. The only man whom I think will knock out whomever he fights is Tyson.
D. The only man whom I think will knock out whoever he fights is Tyson.
E. The only man whom I think will knock out whom he fights is Tyson.

13.
A. The more prettier of my two dresses is the red one.
B. The most pretty of my two dresses is the red one.
C. The prettier of my two dresses is the red one.
D. The prettiest of my two dresses is the red one.
E. The most prettiest of my two dresses is the red one.

14.
A. I told father that a college president was invited to talk.
B. I told Father that a college president was invited to talk.
C. I told father that a College President was invited to talk.
D. I told Father that a College president was invited to talk.
E. I told father that a college's president was invited to talk.

15.
A. One should either be able to read English or Spanish.
B. One should be able either to read Spanish or English.
C. One should be able to either read English or Spanish.
D. One should be able to read either Spanish or English.
E. One should either be able to read either English or Spanish

STATEMENT

When a firefighter that does not have the appropriate training or experience makes a fire company inspection, the caliber of the inspection may not only be substandard, but it may also give the building owner a false sense of security. Only after firefigfhters are trained to identify fire hazards and to know how to correct conditions that have been found are they qualified as competent inspectors.

ACCORDING TO THE PREVIOUS STATEMENT

16. The main idea given is:

A. nearly any firefighter can become a fire inspector
B. fire prevention inspections should not be conducted by untrained firefighters
C. firefighters have to able to identify hazards but not necessarily to correct them
D. substandard fire inspections are only a danger to the fire department

STATEMENT

Foam fire extinguishers use three chemicals, bicarbonate of soda, aluminum sulphate, and a stabilizer. The small interior section will usually be comprised of a 50% water solution of aluminum sulphate. The larger outer section will be comprised of approximately 7% bicarbonate of soda, 3% stabilizer, and 90% water. The stabilizer is used to make the bubbles smaller in diameter and more persistent. When the fire extinguisher is inverted the chemicals of the two sections will intermix, which creates the foam and causes it to expel. The main extinguishing agent is the small bubbles of carbon dioxide gas that will be trapped within the walls of the aluminum hydrate, which forms a hardy, durable, flexible and sticky foam that is able to withstand a considerable amount of abuse. 90% of the foam consists of carbon dioxide gas volume, although approximately 75% of the foam is water by weight. The foam will create a blanket of bubbles over the ignited substance, which will eliminate the air and cool the surface. Cellulose products such as fabric are made fire resistant by the strengthened bubbles containing aluminum hydrate. Both horizontal and vertical surfaces are coated and isolated from the heat by the foam adhering wherever it is utilized including floating over liquid surfaces. The foam is harmless to firefighters. The foam creates less of a wetting effect than water. The extinguisher is not effective on fires involving alcohol. The extinguisher should be protected from low temperatures.

ACCORDING TO THE PREVIOUS STATEMENT:
(answer questions 17 & 18)

17. The stabilizer in a foam fire extinguisher has the primary function of preventing:

A. clinging of the foam to the extinguisher
B. early chemical reaction of the ingredients
C. swift breaking-up of the carbon dioxide bubbles
D. Vaporization of the aluminum sulphate

18. The primary justification for removing the air from an ignited substance with the use of foam is that:

A. fire needs air to continue burning
B. the layer of foam will allow the heat to escape
C. The ignition temperature will be lowered by the foam
D. Heat from the fire will break down the carbon dioxide

19. Reaching the summit appeared beyond the capacity of the hikers. As used in the above sentence, the word summit means most nearly:

A. canyon
B. peak
C. plateau
D. ravine

20. The plot was thwarted by the quick action of the officer. As used in the above sentence, the word thwarted means most nearly:

A. blocked
B. discovered
C. punished
D. solved

21. A gradual decrease in the width of an elongated object is called
A. bevel
B. slope
C. spiral
D. taper

22. The foreman was ordered to expedite production of the vital part. As used in the above sentence, the word expedite means most nearly:

A. accelerate
B. control
C. improve
D. revise

23. Rules must be applied with discretion. As used in the above sentence, the word discretion means most nearly:

A. impartiality
B. judgement
C. severity
D. patience

STATEMENT

A fire of undetermined origin started in the warehouse shed of a flour mill. Although there was some delay in notifying the fire department, they practically succeeded in bringing the fire under control when a series of dust explosions occurred which caused the fire to spread and the main building was destroyed. The fire departments efforts were considerably handicapped because it was undermanned, and the water pressure in the vicinity was inadequate.

ACCORDING TO THE PREVIOUS STATEMENT (questions 24 - 26)

24. From the information in the above paragraph, it is most accurate to state that the cause of the fire was:

A. suspicious
B. unknown
C. accidental
D. arson
E. spontaneous combustion

25. In the fire described above, the fire departments efforts were handicapped chiefly by:

A. poor leadership
B. out-dated apparatus
C. uncooperative company employees
D. insufficient water pressure
E. poorly trained men

26. In the fire described above, the most important cause of the fire spreading to the main building was the:

A. series of dust explosions
B. delay in notifying the fire department
C. inadequate water pressure
D. lack of manpower
E. wooden construction of the building

27. DESIRABLE is to ACCEPTABLE as RESOLVED is to:

A. deleterious
B. determined
C. demure
D. absolved

28. CHASTE is to SENSUOUS as GOOD is to:

A. deviltry B. revel C. candid D. evil

29. FAN is to AIR as NEWSPAPER is to:

A. literature
B. reporter
C. information
D. subscription

30. FUNDAMENTAL means most nearly:

A. adequate
B. essential
C. official
D. truthful

31. BARRICADE means most nearly:

A. bell tower
B. platform
C. barrier
D. lunch counter

32. HAZARDOUS means most nearly:

A. thrilling B. slow C. dangerous D. deceptive

33. INTERSECT means most nearly:

A. run parallel
B. changes names
C. end
D. cross

34. APPROACH means most nearly:

A. discover
B. leave
C. draw near
D. pass through

35. Firefighting ability, in general, is measured in terms of:

A. physical strength
B. ability to follow orders
C. intellectual superiority
D. reasoning ability

36. Firefighters sometimes smash windows and chop holes in roofs in order to:

A. reach trapped victims
B. remove burning materials
C. ventilate a building
D. escaped from a fire

37. Of the following, the chief reason why firefighters may try to avoid wearing breathing apparatus whenever possible is that they tend to:

A. irritate the face
B. be costly
C. be unsanitary
D. hamper working ability

38. Suppose that you are a member of an engine company which was notified to respond to a fire near Box 8888. During the response, your engine company encounters another fire in a warehouse. For your engine company to stop to extinguish the warehouse fire would be imprudent chiefly because:

A. warehouses are well protected with sprinkler systems
B. your company may be the only engine company notified to respond to the fire near Box 8888
C. other engine companies may also be responding to the fire near Box 8888
D, time would be wasted in summoning another engine company to the warehouse fire.

39. In manufacturing plants, it is often better to allow smoking in safe places and at specified hours when supervision can be maintained than to run the risk of surreptitious smoking in out-of-the-way places. The previous statement illustrates the point that:

A. an ounce of prevention is worth a pound of cure
B. it is not inconsistent with acceptable safety procedure to regulate factory smoking
C. smoking should always be supervised
D. factory fire caused by surreptitious smoking is generally a lesser fire hazard than one starting when smoking is regulated

40. When the oxygen concentration of air is less than 16%:

A. breathing is not affected
B. most substances will continue to burn freely
C. most flames are extinguished
D. no fire will continue burning

41. The two main factors which determine a fire extinguishers worth are:

A. classification and size
B. accessibility and serviceability
C. extinguishing agent and volume
D. weight and ease of operation

42. Electrical insulators are usually made of:

A. copper
B. brass
C. porcelain
D. steel

43. Training for a firefighter ends:

A. when he completes recruit training
B. when he has been promoted to his highest level
C. when he becomes a chief officer
D. none of the above; training never ends for firefighters

44. Of the following, the chief advantage likely to be derived from installing a short-wave radio receiving set in every piece of fire apparatus is that:

A. fire companies could be communicated with while they are in the field
B. traffic accidents involving fire department apparatus would be reduced
C. fewer men would require technical fire training
D. a definite wave band could be set aside for exclusive use by the fire department

45. A firefighter on duty who answers a department telephone should give his name and rank/position:

A. at the start of the conversation, as a matter of routine
B. only if asked for this information by the caller
C. only if the caller is a superior officer
D. only if the telephone message requires the firefighter to take some action

46. Of the following incipient fires, which one would water fire extinguishers be **LEAST** effective on:

A. textiles
B. paper
C. wood
D. greases in open containers

47. The most important reason for having firefighters employed by the fire department wear uniforms is to:

A. indicate the semi-military nature of the fire department
B. build morale and esprit de corps of firefighters
C. identify firefighters on duty to the public and other firefighters
D. provide clothing suitable for the work performed

48. 12,578 rounded off to the nearest hundred is:

A. 13,000
B. 12,600
C. 12,570
D. 12,500

49. The largest part of the brain is called the:

A. medulla
B. cortex
C. cerebrum
D. cerebellum

50. When responding to alarms, the fire department apparatus generally follows routes established in advance. The one of the following which would be the least valid justification for this practice is:

A. motorists living in the area become familiar with these routes and tend to avoid them
B. the likelihood of collision between two pieces of fire department apparatus is reduced
C. the fastest response generally is obtained
D. road construction, road blocks, detours and similar conditions can be avoided

51. Why is it beneficial to enter a smoke filled room crawling low to the floor?

A. smoke will be radiated through the floor
B. smoke is lighter than air
C. smoke may become flame
D. smoke will be compressed in a room

52. About 78% of the air exhaled by a person is:

A. carbon dioxide
B. nitrogen
C. water vapor
D. hydrogen

53. If you are in your home at the time of an atomic bomb attack, the best procedure to follow would be to:

A. start running to the nearest air raid shelter center
B. turn on the radio and wait for instructions
C. prepare to evacuate the area
D. lie down in the most sheltered place, away from windows

54. When a material is referred to as combustible, it most likely is:

A. fireproof
B. flammable
C. crammed
D. disintegrated

55. Spontaneous combustion may be the reason for a pile of oily rags catching fire. In general, spontaneous combustion is the direct result of:

A. application of flame
B. falling sparks
C. intense sunlight
D. chemical action

56. When the gears of an automobile are shifted from high to low:

A. the power of the motor is increased
B. the force applied by the driving wheels is increased
C. the force applied by the motor is decreased
D. the speed of the driving wheels is increased

57. Hard coal burns with less smoke than soft coal because hard coal:

A. is more nearly pure carbon
B. contains more volatile materials
C. has a lower kindling temperature
D. undergoes chemical change more readily

58. The liquid in an automobile battery which is in good condition is:

A. an acid solution
B. a salt water solution
C. a caustic solution
D. water only

59. To say that something is "soluble" in water means most nearly that it:

A. floats
B. sinks
C. dissolves
D. crumbles

60. Water in the cooling system of automatic fire apparatus should be:

A. alkaline
B. acid
C. salty
D. neutral

STATEMENT

Magnesium combines so readily with oxygen that under some conditions water applied to extinguish magnesium fires may be decomposed into its constituent elements, oxygen and hydrogen. The oxygen combines with the magnesium and the hydrogen released adds to the intensity of the fire.

ACCORDING TO THE PREVIOUS STATEMENT

61.

A. oxygen in water will combine with the magnesium to extinguish a magnesium fire
B. when water decomposes into its elements, its effectiveness in fighting magnesium fires is increased
C. water applied to a magnesium fire may even increase the fire
D. a magnesium fire may be expected to react to water in the same way as most other light metals
E. water is useless in fighting magnesium fires

62. Which of the following is not regarded as a primary hazard in a fire:

A. formation of carbon dioxide
B. heat
C. oxygen depletion
D. formation of carbon monoxide

63. Which one of the following is a product of combustion:

A. dust
B. friction
C. flame
D. oxygen

64. Find percentage:

2% of $18.64 = (express to nearest cent)

A. $3.74 B. $.37 C. $.38 D. $14.95

65. Add:

18 + 16 + 9 =

A. 39 B. 43 C. 49 D. 53

66. Divide:

76,500 by 90 =

A. 535 B. 540 C. 545 D. 570

67. Multiply :

843 X 90 =

A. 75,870 B. 75,970 C. 75,720 D. 7,587

68. Add:

54,261 + 2,271 + 511 + 13,462 =

A 70,505 B. 70,515 C. 70,605 D. 70,555

69. Find the percentage:

20% of 30 =

A. 5 B. 15 C. 3 D. none of these

70. How many cubic feet of water will a tank on the back of a truck whose length is 8ft, width is 5ft, and height is 3 1/2 ft hold?

A. 100 cubic feet
B. 130 cubic feet
C. 140 cubic feet
D. 150 cubic feet

71. A chain-saw cuts 8 logs per minute. If there are 1,440 logs to be cut, the number of hours required to cut these logs is:

A. 2 1/2 B. 2 2/3 C. 4 D. 3 [illegible]

72. Firefighter Simpson must check the supply of air tank cylinders at the beginning of each tour of duty. There must be ten air tank cylinders always full of air and ready to be exchanged for used, empty air tank cylinders. At the start of a new tour of duty, Firefighter Simpson finds that out of twenty cylinders present, only five cylinders are full and ready to be exchanged. What is the minimum number of used empty cylinders that Firefighter Simpson must replace with full cylinders?

A. 5 B. 10 C. 15 D. 20

73. Subtract 2 hours and 50 minutes from 5 hours and 12 minutes

A. 3 hrs 38 minutes
B. 2 hrs 22 mins
C. 2 hrs 32 mins
D. 3 hrs 22 mins

74. Write the figure: eight and one-half million:

A. 8,500,000
B. 850,000
C. 8,500
D. 8,500,000,000

75. Fire Captain Jones stated to Firefighter Smith that light travels approximately 186,000 miles per second and that the sun is 933 million miles away from the earth. Firefighter Smith asked Captain Jones if he could figure out how long it would take a ray of light to travel from the sun to the earth. Captain Jones said he could figure it out to the nearest minute. What would be the correct time?

A. 2 minutes B. 5 minutes C. 8 minutes D. 58 minutes

IN THE FOLLOWING DRAWINGS, CHOOSE THE ARTICLE FROM THE RIGHT THAT MAY BE MADE FROM THE FLAT DESIGN ON THE LEFT (questions 76 - 79)

76.

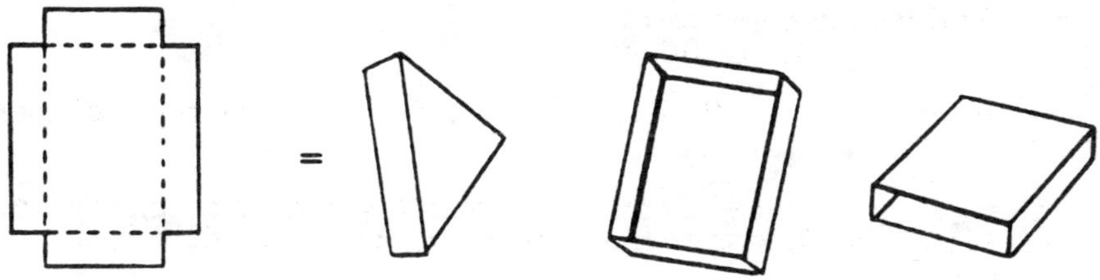

DESIGN A. B. C.

77.

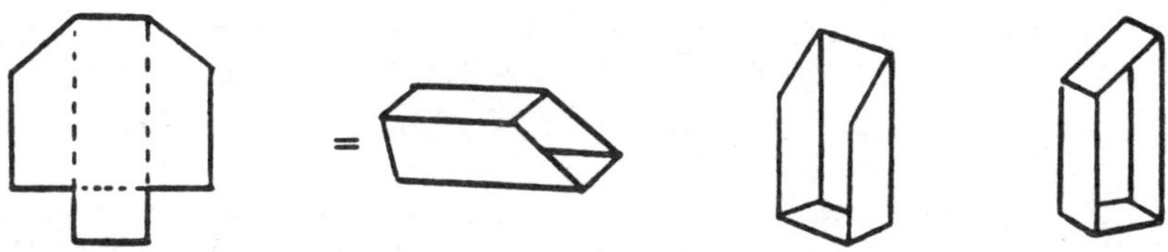

DESIGN A. B. C.

78.

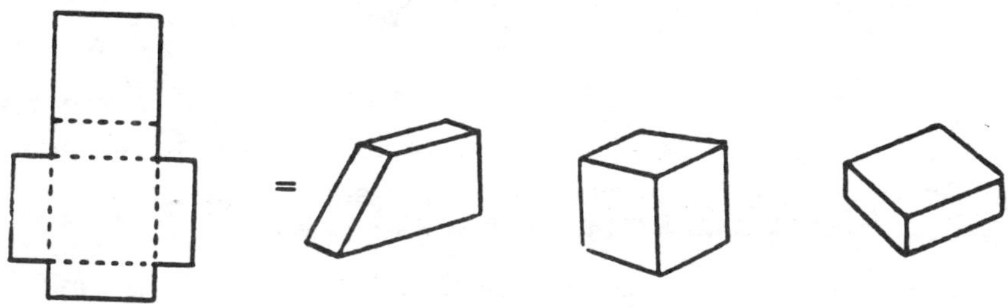

DESIGN A. B. C.

79.

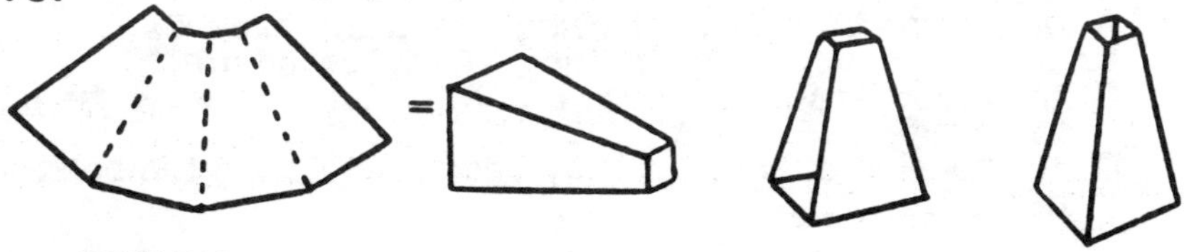

DESIGN A. B. C.

FIND THE LETTER WHICH COMES NEXT IN THE FOLLOWING ALPHABETICAL PROGRESSIONS (questions 80 & 81)

80.
G H I J K __

A. L B. M C. O D. P

81.
XX UU RR ___

A. OO B. PP C. QQ D. RR

FIND THE NUMBER WHICH COMES NEXT IN THE FOLLOWING NUMERICAL PROGRESSIONS (questions 82 & 83)

82.
25 27 30 15 17 20 10 ___

A. 8 B. 12 C. 25 D. 30

83.
4 6 12 14 28 30 60 62 ___

A. 64 B. 68 C. 124 D. 128

84. The breakdown of the machine was due to a defective gasket. As used in the above sentence, the word gasket means:

A. filter B. piston C. sealer D. transmission

85. The tool that would be most appropriate to cut a chain securing a gate on a fence would be a/an:

A. axe B. bolt cutter C. pry bar D. tin snip

86. A clay pitcher containing water will crack if the water freezes because:

A. during the process of freezing, the water expands
B. during the process of freezing, the water contracts
C. the crystallization process of ice formation causes the ice to solidify
D. the tension strength of clay decreases as the temperature decreases

87. The one of the following statements about electric fuses that is most valid is that they:

A. should never be replaced by coins
B. may be replaced by coins for a short time if there are no fuses available
C. may be replaced by coins provided that the electric company is notified
D. may be replaced by coins provided that care is taken to avoid overloading the circuit
E. may be replaced only by a licensed electrician

88. The object shown below is used to:

A. refill batteries
B. extinguish fires
C. lubricate vehicles
D. kill insects

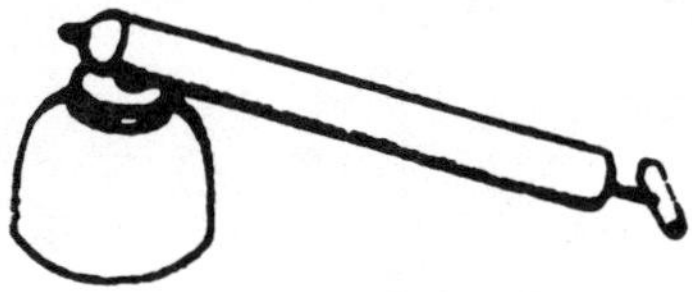

89. The object shown below is used to:

A. scrape paint
B. etch leather
C. install window glass
D. drive screws in close areas

90. The object shown below is a:

A. manifold gasket
B. brake band
C. clutch plate
D. air filter

91. Of the two following drawings, which one has the weights which will move?

A. drawing #1
B. drawing #2
C. drawings #1 and #2
D. neither drawing

92. In the drawing below, the turnbuckle should be turned in which direction in order to shorten it:

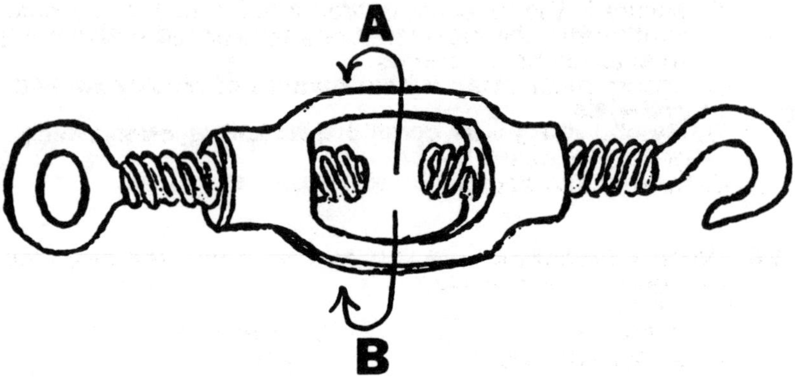

A. direction A
B. Direction B
C. it would not make any difference

93. In the drawing below, if #1 and #2 are of equal size and weight, and the pulley at the top will turn freely, which one of the pulleys will tend to roll down the ramp?

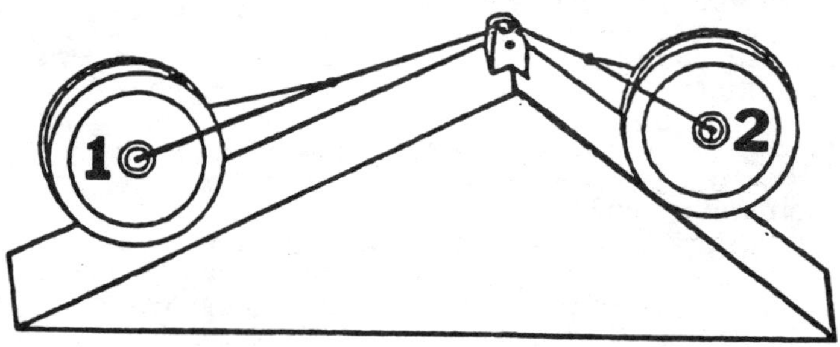

A. #1　　B. #2　　C. it will not make a difference

94. The one weakness offered by windows of wired glass is that:

A. the wire mesh is a conductor of heat
B. they shatter when struck by a cold stream
C. they radiate heat
D. they disintegrate in a severe fire

95. Spontaneous heating and ignition:

A. has many fundamental causes
B. occurs if the inherent characteristics of the material cause exothermic chemical reactions to proceed without exposure to external heat sources
C. occurs most often within masses of closely packed materials
D. usually starts with a rapid chemical reaction which generates heat
E. none of the above

96. When a firefighter suffers a burn at a fire, the direct cause of this burn is most likely:

A. flame
B. carbon monoxide
C. smoke
D. heat

97. Normally the first victims to be rescued during a search are those who are:

A. unconscious
B. trapped under rubble
C. easy to see and hear
D. injured

98. A muffler is used in an automobile to:

A. keep the passengers warm
B. diminish noise created by waste gases
C. remove poisonous carbon monoxide
D. keep the radiator from freezing
E. none of these

99. The basic assumption of fire prevention educational programs is that people frequently:

A. must be forced into obeying fire laws
B. are unaware of the dangers involved in some of their actions
C. don't care whether or not their actions are dangerous
D. assume that fire insurance protects them against all fire loss

100 When the inside face of a rubber suction cup is pressed against a smooth wall, it usually remains in place because of the:

A. force of molecular attraction between the rubber and the wall
B. pressure of the air on the rubber cup
C. suction caused by the elasticity of the rubber
D. static electricity generated by friction between rubber and the wall

PRACTICE EXAM #5
ANSWERS

1. A
2. C
3. D
4. A
5. A
6. C
7. A
8. B
9. B
10. A
11. B
12. B
13. C
14. A
15. D
16. B
17. C
18. A
19. B
20. A
21. D
22. A
23. B
24. B
25. D
26. A
27. B
28. D
29. C
30. B
31. C
32. C
33. D
34. C
35. B
36. C
37. D
38. B
39. C
40. C
41. B
42. C
43. D
44. A
45. A
46. D
47. C
48. B
49. C
50. A
51. B
52. B
53. D
54. B
55. D
56. B
57. A
58. A
59. C
60. D
61. C
62. A
63. C
64. B
65. B
66. A
67. A
68. A
69. D
70. C
71. BD
72. A
73. B
74. A
75. C
76. B
77. B
78. C
79. C
80. A
81. A
82. B
83. C
84. C
85. B
86. A
87. A
88. D
89. D
90. C
91. D
92. A
93. B
94. C
95. B
96. D
97. C
98. B
99. B
100. B

PRACTICE EXAM
#6

FIREFIGHTER WRITTEN

PRACTICE EXAM #6

1. Some tools are known as all-purpose tools because they may be used for a great variety of purposes; others are called special-purpose tools because they are suitable only for a particular purpose. Generally, an all-purpose tool, compared to a special tool for the same purpose, is:

   A. cheaper
   B. less efficient
   C. safer to use.
   D. simpler to operate

2. Physical fitness is not the only requirement of the successful firefighter. His moral character must be such as to withstand the most severe tests. His word must be as good as his bond. According to these statements, a firefighter must:

   A. have judgement enough to handle difficult situations
   B. be as intelligent as he is physically fit
   C. give a bond before he is appointed
   D. be trustworthy in every respect

3. Of the following choices, reasonably the primary requisite for safe working conditions would be:

   A. smart workmanship
   B. modern equipment
   C. good weather
   D. good housekeeping

4. "Smoke" is a visible product of incomplete combustion. Smoke ordinarily encountered in a fire consists of a mixture of oxygen, nitrogen, carbon dioxide and some carbon monoxide, finely divided particles of soot and carbon, and a miscellaneous assortment of products which have been released from the material involved. In a burning structure, smoke builds up gradually and continuously reduces visibility until ventilation is accomplished. Lack of visibility causes disorientation which can trap persons in a smoke-filled building.

   According to the previous statement: The more common combustible materials contain:

   A. carbon
   B. gases
   C. oxygen
   D. carbon monoxide

5. Which one of the following is considered to be a product of combustion when a fuel burns:

A. electrical energy
B. oxygen
C. gasoline
D. fire gases

6. OIL is related to GREASE as WAX is related to:

A. grit B. bee C. paraffin D. relish

7. CAR is related to BREAKS as SHIP is related to:

A. anchor B. engine C. captain D. ocean

8. LOOK is related to SEE as LISTEN is related to:

A. consider B. refrain C. hear D. observe

9. SCHOOL is related to TEACHER as HOSPITAL is related to:

A. tragedy B. comedy C. physician D. operation

10. In the following group of words, which is the word that should NOT be in this group of words.

A. insulation B. convection C. radiation D. conduction

11. When a firefighter must make a decision and the only choices are both unsatisfactory courses of action, this situation is best described as a:

A. farce B. dilemma C. crisis D. deadlock

12. To say that a man meets all of the requirements for appointment as a fireman means that he is:

A. assigned B. eligible C. elected D. hired

13. Several firefighters were injured by the detonation. As used in this sentence , the word detonation means most nearly:

A. accident B. collapse C. collision D. explosion

14. "Lawer" is related to "Client" as "Doctor" is related to:

A. baby B. patient C. mother D. insurance

15. INNOVATION means most nearly:

A. change B. prayer C. hint D. restraint

16. COMPILE means most nearly:

A. confuse B. support C. compare D. gather

17. MANUALLY means most nearly:

A. by man
B. manufactured
C. several ways
D. by hand
E. in the dark

18. ELABORATE means most nearly:

A. wealthy
B. praiseworthy
C. structural
D. modified
E. detailed

19. PRESCRIBE means most nearly:

A. forbid
B. picture
C. dictate
D. underwrite
E. foresee

20. REVISE means most nearly:

A. introduce
B. explain
C. begin
D. change
E. accept

CHOOSE THE CORRECT SPELLING OF THE WORD IN THE FOLLOWING QUESTIONS: (questions #21 & #22)

21.

A. fundamental
B. fundemental
C. fundamentol
D. none of the above

22.

A. frictcion
B. frction
C. friction
D. none of the above

23. Which of the following words is misspelled?

A. asphixiate
B. resuscitate
C. arsonist
D. extension

24. Which of the following words is spelled correctly?

A. celler
B. acidently
C. discipline
D. escalater

25. Which of the following words is misspelled?

A. catridge
B. combustion
C. convection
D. inhalator

26. Which of the following words is misspelled?

A. immagine
B. viaduct
C. chisel
D. Saturn

SELECT THE SENTENCE WHICH CONTAINS AN ERROR IN GRAMMAR, USAGE, OR PUNCTUATION (questions 27 - 30)

27.
A. Yes, we have all kinds of assorted pies.
B. His fatigue was due to the long march.
C. Swathed in bandages from head to foot, he was a pitiable sight.
D. Since we studied the case together, we have had a common interest in the child.
E. He was graduated from Seton Hall University.

28.
A. We saw the man, who in spite of the orders not to cross the police lines, ran into the burning.
B. I have offered to forego this year's interest; I will make no further offers.
C. Despite his insistence on attending the meeting, I repeat that he will not go.
D. The injured man lay there for two days before he was found.
E. I do not remember ever having met him.

29.
A. Both the classical and the general course prepares the student for college.
B. He had grown tall and much heavier.
C. The reason he refuses to face the issue is that he is a coward.
D. He left, didn't he?
E. In that he offered no protest, he was as responsible for the child's injury as those who inflicted it.

30.
A. After spending several weeks there, he wrote that he was enchanted with the countryside.
B. "Who's willing to following me?" he asked.
C. Whom do you wish to appoint to head the committee?
D. The most important part of the whole mechanism is the ball bearings.
E. Neither of the boys was willing to go.

31. If you were to notice flames coming out of a factory window situated on a lonely street, the first thing you should do is:

A. phone the police
B. call the operator and tell her you want to report a fire
C. see whether the fire is under control
D. send in a fire alarm

32. Two firefighters, on their way to report for duty early one morning, observe a fire in a building containing a supermarket on the street level and apartments on the upper stories. One firefighter runs into the building to spread the alarm to the tenants. The other firefighter runs to a street alarm box two blocks away and sends an alarm. The latter fireman should then:

A. return to the building which is on fire and help evacuate the tenants
B. remain at the fire alarm box in order to direct the first fire company that arrives at the location of the fire
C. look for a telephone in order to call his own fire company and explain that he and his companion will be late in reporting for duty
D. look for a telephone in order to call the health department and request that an inspector be sent to the supermarket to examine the food involved in the fire

33. Of the following fire extinguishers, which one would be most useful on an electrical fire?

A. foam
B. water
C. dry chemical
D. vaporizing liquid

34. Firefighters should remember that the part of the rope that is to be used for hoisting, pulling, snubbing, etc. is known as the:

A. standing part
B. running part
C. working part
D. tension section

35. A firefighter in uniform, performing a fire inspection, notices a throng of "gang-bangers" attacking a police-officer. The firefighter goes to the aid of the police-officer and, during the fight, he obtains minor injuries. The action of the firefighter in this situation was:

A. proper, chiefly because members of the uniformed forces of a community must stick together
B. improper, chiefly because people in the neighborhood, as a result, might refuse to cooperate with the fire departments various programs
C. proper, chiefly because all citizens have an obligation to assist policemen in the performance of their duty
D. improper, chiefly because the fire department lost the services of the firefighter while he was recovering from his injuries

36. When a firefighter is assigned a rush job, normally the best way to complete the assignment on time is to:

A. first look the task over and organize it
B. start to work on the assignment immediately so that no time is lost
C. finish current assignments already on hand so that it will not be in the way
D. organize the work area so that work can be performed efficiently

37. Generally, company fire inspections will not include the living quarters of private dwellings unless the occupants request the inspection. Of the following explanations as to why private dwellings are excluded from mandatory inspection is that:

A. private dwellings rarely catch fire
B. private dwelling fires are normally extinguished easier than other types of fires
C. people may take offense at such inspections as an intrusion of privacy
D. the dollar value of private dwellings is lower than that of other types of occupancies

38. In a room filled with smoke the best air is located near the:

A. ceiling B. windows C. floors D. doors

39. Fire inspectors are aware that the most common method of operation for fire doors is:

A. the single-slide installation
B. swinging doors
C. telescoping vertically gliding doors
D. rolling doors

40. Grass and weeds should not be permitted to grow high near structures, primarily because, if a fire should occur, the tall weeds and grass may:

A. emit toxic fumes
B. restrict maneuverability of firefighters
C. disrupt with the egress of occupants from the structure
D. extend the fire to the structure and set it on fire

41. The exterior wall vertical studs are continuous from the foundation to the roof. The joists for the first floor are nailed directly to the studs without a header. Firefighters should recognize that the previous statement describes what type of wall construction?

A. platform frame
B. balloon frame
C. western frame
D. tilt-up construction

42. Firefighters should be aware that many fires are caused by inappropriate use of oxy-acetylene torches. The principal cause of these fires is:

A. the elevated pressure under which the gases are stored
B. failure to contain or extinguish sparks
C. high temperatures produced by the apparatus
D. explosive character of the gases utilized

43. Firefighters should know that of the apparatus listed below, the apparatus that is not equipped to provide elevated streams is:

A. aerial ladder
B. aerial ladder tower
C. telescoping-aerial tower
D. articulating-aerial tower

44. An on duty firefighter at an amphitheater realizes that spectators standing in the aisles are obstructing the aisles. The firefighter without delay should:

A. relate the condition to his/her superior officer
B. direct the spectators to relocate out of the aisles
C. ask the manager of amphitheater to correct the predicament
D. give a written violation notice to the manager of the amphitheater

45. The accumulation of hot gases are most dangerous when they are in an area that is:

A. large B. closed C. damp D. cool

46. Spontaneous combustion can be the cause of ignition to a a stack of oily rags. Generally, spontaneous combustion is the direct result of:

A. use of flame
B. chemical action
C. profusion of sparks
D. radioactivity

47. Boiling water put into a glass, probably will crack the glass. When boiling water is put into a glass that has a metal object, such as a spoon, located within the glass, the glass is much less likely to crack. The reason that the glass with the metal object is less likely to crack is that the metal object:

A. disperses the water over a larger surface area of the glass
B. promptly absorbs the heat from the water
C. strengthens the glass
D. decreases the amount of water which can be put into the glass

48. Of the following, the most significant reason why firefighters should wear fire-helmets is that the helmets will:

A. shield the firefighters head from heat
B. set the diverse fire companies out from each other
C. protect firefighters from falling objects
D. keep water off the firefighters heads

49. Firefighters should be aware that petroleum is:

A. manufactured from gasoline
B. employed only as a lubricant
C. the major source of coke
D. lighter than water
E. utilized only as a fuel

50. The point at which combustion takes place is generally known as the:

A. boiling point
B. combustion point
C. flash point
D. ignition point

51. The yellow flame in a kerosene lamp is caused by the:

A. burning of hydrogen
B. burning of the wick
C. thorough burning of the hydrocarbons
D. illumination of unburned carbon particles

52. Firefighters should be aware that the reason why smoke usually rises from a fire is that:

A. cooler, heavier air displaces lighter, warm air
B. heat energy of the fire drives the smoke upward
C. suction from the upper air draws the smoke upward
D. burning material is chemically converted into heat energy

53. When a mass of air rises, it is cooled primarily because:

A. it expands
B. it combines with cold air
C. it is closer to cold outer space
D. ice crystals are developed

54. Oil value as a lubricant depends upon its:

A. mixture B. origin C. viscosity D. color

55. An electron is a/an:

A. electronic nucleus
B. hydrogen atom
C. static charge
D. particle of negative electricity

56. Firefighters should be aware that _________ is not a necessary element of fire?

A. draft B. oxygen C. fuel D. heat

57. Assume that two duplicate insulated water tanks are filled with equal amounts of faucet water. A large piece of ice is placed in one tank and the same quantity of ice, chopped into small pieces, is placed in the other tank. Which of the following reasons is the most likely justification that the water in the tank containing the chopped ice, compared to the water in the other tank, will be chilled:

A. faster but to a substantially higher temperature
B. faster and to approximately the same temperature
C. slower but to a substantially higher temperature
D. slower and to the same temperature

58. The reason that when a standing subway passenger, holding onto a safety hand "tie", pulls the METAL safety hand "Tie", the "tie" swings toward the passenger and when it is released, it swings back to it's original position. The best explanation of why the handle returns to it's original position is that it is:

A. counter-weighted
B. responding to centrifugal forces acting upon it
C. spring actuated
D. well-balanced

59. The largest planet of our solar system is:

A. Mars B. Earth C. Jupiter D. Saturn

60. Which of the following would represent the greatest threat to an explosion:

A. square tanks
B. round tanks
C. full tanks
D. almost empty tanks

61. Firefighters are apprised of the fact that most gases produced in fires are toxic. Fire fatalities are generally due to the inhalation of fire gases and hot products of combustion rather than to direct burns.

ACCORDING TO THE ABOVE STATEMENT

A. fire gases created during a fire normally are harmful if inhaled into the lungs
B. fire gases are most hazardous if permitted to come in contact with the skin
C. more fire injuries are produced by the fire itself than by fire products
D. most fire gases produced during a fire are more harmful to the burning building than to firefighters
E. most of the fire gases which are existing in burning structures are nontoxic except when they are burning

62. During a fire incident THIS product of combustion, normally encountered, includes a blend of oxygen, nitrogen, carbon dioxide, some carbon monoxide, and small particles of soot and carbon. THIS product of combustion is known as:

A. flame
B. smoke
C. heat
D. carbon monoxide

63. If "y" represents the tens digit and "x" the units digit of a two digit number, then the number is represented by:

A. y + x B yx C. 10x + y D. 10y + x

64. The City Fire Department baseball team won 4 games and lost 20 games. what fraction of its games did the team win?

A. 1/4 B. 1/5 C. 4/5 D. none of these

65. A truck company needs to cover a ventilation trench 18 feet long and 1 yard wide. They have enough lumber is to cover two-thirds (2/3) of the trench area. How many square feet will remain uncovered?

A. 2 square feet
B. 4 square feet
C. 12 square feet
D. none of these

66. A flashlight cost a firefighter $72.00 which includes batteries. The firefighter wants to sell the flashlight to another firefighter at a 20 percent profit. How much should the firefighter sell the flashlight for?

A. $88.00 B. $90.00 C. $92.00 D. $130.00

67. Multiply:

64 X 3 =

A. 182 B. 188 C. 192 D. 202

68. 468 divided by 6 =

A. 71 B. 73 C. 74 D. 78

69. Find the percentage (express to the nearest cent):

18% of $134.15 =

A. $24.15 B. $24.14 C. $2.41 D. $158.30

70. Add:

13,589 + 29,462 + 94,001 + 87,503 =

A. 224,955 B. 224,555 C. 204,595 D. 224,556

71. Multiply:

1,826 X 1039 =

A. 253,914 B. 1,897,175 C. 253,916 D. 1,897,214

72. A cylindrical storage tank has a radius of 3 1/2 feet. Its height is 8 feet. Using pi = 22/7, find the volume in cubic feet

A. 300 cubic feet
B. 304 cubic feet
C. 308 cubic feet
D. 312 cubic feet

73. Convert 25 2/3 to an improper fraction of thirds:

A. 75/3 B. 77/3 C. 81/3 D. 88/3

74. One half dozen road flares added to four road flares will equal how many road flares?

A. 9 B. 10 C. 12 D. 16

75. Firefighter Derouin had three library books that were 4 days overdue. His fine was $.12 per day for each book. How much was his total fine?

A. $1.24 B. $1.34 C. $1.43 D. $1.44

INSERT THE CORRECT LETTER THAT WILL COMPLETE THE FOLLOWING ALPHABETICAL PROGRESSIONS:

76.
M J G D __

A. N B. M C. P D. R

77.
A C B D C E __

A. A B. B C. C D. D

78.
Z W T __

A. Q B. R C. S D.U

IN THE FOLLOWING NUMBER PROGRESSIONS ADD THE CORRECT NUMBER THAT SHOULD BE PLACED AT THE END OF EACH SERIES:

79.
1 3 5 7 9 11 13 __

A. 10 B. 12 C. 14 D. 15

80.
20 10 12 6 4 2 4 __

A. 3 B. 2 C. 1 D. 0

FIND THE LETTER WHICH COMES NEXT IN THE FOLLOWING ALPHABETICAL PROGRESSIONS:

81.
A D G J ___

A. L B. N C. I D. M

82.
Z W T Q ___

A. P B. R C. N D. O

83.
B D G K ___

A. N B. P C. Q D. M

84. Of the following list of tools, which is the one that would be the BEST to use when a precise measurement of the inside diameter of a cylindrical object is needed

A. depth gauge.
B. vernier calipers.
C. micrometer calipers
D. measuring tape

85. The use of FLUX when soldering to conductor wires together is to:

A. prevent oxidation from heat
B. create a brilliance on the surface
C. prevent the connection from overheating
D. keep a constant temperature on the soldering tool

86. The chief benefit of a four wheel driven vehicle has over a normal vehicle is:

A. increased speed.
B. better fuel consumption
C. increased traction
D. greater durability.

87. Firefighters should recognize that the flying buttress is most closely affiliated with:

A. fire fighting
B. combat
C. architecture
D. flying
E. forestry

88. As the elevation above sea-level increases, the atmospheric pressure decreases at the rate of approximately:

A. one-half pound for every 500 feet of elevation
B. one-half pound for every 1000 feet of elevation
C. one pound for every 3000 feet of elevation
D. two pounds for every 1000 feet of elevation
E. three pounds for every 2000 feet of elevation

89. Fire inspectors should recognize that glass block panels are regarded as:

A. appropriate for light exposures
B. to have relatively low thermal insulation
C. proper for moderate exposures
D. suitable for use as bearing walls

90. The primary objective of a trap in a plumbing fixture is to:

A. prevent any material from obstructing the system
B. prevent the passage of gas or air through the fixture
C. allow the free transition of liquids
D. add exterior frill and finish to fixtures
E. expand the rate of discharge

91. A tool that is used to loosen or tighten a set screw is sometimes called a hex key or set-screw wrench. This tool is also known as:

A. a crocodile wrench
B. a Stillson wrench
C. an Allen wrench
D. a crescent wrench

92. Firefighters frequently cut ventilation holes in roofs in order to allow smoke and fire gases to escape from burning structures. When cutting these holes the firefighters attempt to place their axe-cuts just along side the heavy supporting beams underneath the planking. Which of the reasons listed below is the reason for positioning the cut just to the side of such beams is that:

A. the planking is thinnest in this area
B. nails are not usually in this area
C. The axe is less likely to bounce in this area
D. the beam provides a strong brace for the firefighter that is using the axe

93. Firefighters should be aware that t takes more energy to force water through a long pipe than through a short pipe of the same diameter, because of:

A. gravity B. friction C. inertia D. cohesion

94. When water is travelling from a fire stream, and is directed at the roof of a three story building, at what point would the water be travelling at its greatest speed?

A. as it falls on the roof
B. at its maximum height
C. at a midway point between the ground and the roof
D. as it leaves the hose nozzle

95. When starting a vehicle equipped with a manual choke on a cold day, it is best to pull the choke out because it:

A. increases the amount of air in the carburetor
B. reduces the amount of fuel entering the carburetor
C. allows more fuel to enter the carburetor for a richer fuel mixture
D. speeds up the supply of air and fuel to the engine

96. The object shown below is a:

A. paper clip
B. wire staple
C. cotter pin
D. finishing pin

97. The object shown below is used to:

A. punch holes
B. clamp objects
C. measure thickness
D. measure depth

98. Firefighters are trained to give artificial respiration mainly to increase:

A. the oxidation of the brain
B. blood circulation to the brain
C. the interchange of air in the lungs
D. blood circulation to the lungs

99. Of the following choices of First Aid for a victim of a head injury, which is correct?

A. have patient lying down with head lower than the rest of the body
B. have patient lying down with head somewhat raised
C. have patient sit with head bent down
D. have patient continue to move head in motion to increase circulation

100 Firefighters should be aware that when searching for buried victims it is very vital that they:

A. have intervals of silence
B. continue to move at all times
C. investigate destruction quickly
D. stay away from piles of debris

PRACTICE EXAM #6
ANSWERS

1. B	26. A	51. D	76. A
2. D	27. A	52. A	77. D
3. B	28. D	53. A	78. A
4. A	29. A	54. C	79. D
5. D	30. A	55. D	80. B
6. C	31. D	56. A	81. C D
7. A	32. B	57. B	82. B C
8. C	33. C	58. C	83. B
9. C	34. A	59. C	84. C
10. A	35. C	60. D	85. A
11. B	36. A	61. A	86. C
12. B	37. C	62. B	87. C
13. D	38. C	63. D	88. B
14. B	39. A	64. D	89. A
15. A	40. D	65. D	90. B
16. D	41. B	66. B	91. C
17. D	42. B	67. C	92. C
18. E	43. A	68. D	93. B
19. C	44. C	69. A	94. D
20. D	45. B	70. B	95. C
21. A	46. B	71. D	96. C
22. C	47. B	72. C	97. C
23. A	48. C	73. B	98. C
24. C	49. D	74. B	99. B
25. A	50. D	75. D	100. A

PRACTICE EXAM

#7

FIREFIGHTER WRITTEN
PRACTICE EXAM #7

1. While shopping at the mall, a citizen discovers an official Fire Department badge and gives it to a nephew as a play thing. What is the main argument as to why the citizen should not have responded in this manner?

   A. It would not be a respectable thing to do.
   B. Children may be injured playing with badges.
   C. The badge should be returned to the Fire Department.
   D. Badges are for adults only.

2. Firefighters should not give lectures concerning the Fire Department without the permission of the Fire Chief. The reason for this is:

   A. To limit the chances of the citizens getting information concerning the Fire Department that may not be correct.
   B. To find out if the Firefighter is involved in activities that are prohibited by the Fire Department.
   C. So the Fire Chief will be aware of Firefighters who are public speakers.
   D. So the Fire Chief will be aware of the off duty activities of the Firefighters.

3. It would not be a good practice to allow newspapers to accumulate in a woodworking business because they may:

   A. Obstruct egress during a fire.
   B. Spontaneously ignite.
   C. Obstruct firefighting operations.
   D. Ignite from a spark or discarded match.

4. A Firefighter's mental attitude is of vital importance because:

   A. His/her efficiency on the job is always visible to the public.
   B. He/she must keep calm in order to think clearly in emergencies.
   C. Firefighting often calls for executive ability.
   D. Intelligence as well as physical stamina is needed.

5. All but one of the following are elements required by a fire, which one is not required:

   A. Fuel.
   B. Heat.
   C. Inhibited chain reaction.
   D. Unhibited chain reaction.

6. An incipient fire is a fire that is:

A. In the beginning phase.
B. An arson fire.
C. A rapidly spreading fire.
D. A fire about to begin.

7. Which of the following best describes the purpose of fire doors

A. Prevent panic.
B. Prevent fires.
C. Prevent arson.
D. Confine fire to a specific location of a structure.

8. Of the following which is the principal requirement for safe working conditions:

A. Housekeeping.
B. Workmanship.
C. Weather.
D. Equipment.

9. Which of the following is the best reason for having Firefighters under competitive civil service:

A. Consolidated liability.
B. Not as many emergency incidents develop.
C. A higher level of efficiency is achieve.
D. Company fire prevention is developed.

10. Of the choices below, which one indicates why the regular cleaning of fire apparatus is necessary:

A. Cleaning minimizes deterioration.
B. Cleaning is a beneficial habit to develop.
C. Cleaning reduces spontaneous combustion.
D. Cleaning fire apparatus demonstrates Firefighters awareness of responsibility.

11. Of the choices below, which describes the reason for a Fire Department having organized "drills" for Firefighters:

A. To demonstrate the Fire Department's fire apparatus to the citizens.
B. To increase and maintain the Firefighter's firefighting skills.
C. To contribute to the physical ability of the Firefighters.
D. To illustrate the Fire Department's efficiency.

12. When a street is referred to as "Arterial", it is a street that:

A. Goes through a city diagonally.
B. Large one-way streets.
C. Intersects a business districts.
D. Reserved for firefighting apparatus during emergencies.

13. As a Firefighter you may have specific information that you instructed not to make public. If a citizen asks you for this information, you should:

A. Asks you Captain if you can tell the citizen.
B. Tell the citizen that the information is only for Fire Dept use.
C. Tell the citizen that you do not feel it would be a good idea for you to give the information to the public.
D. Advise the citizen that you are not allowed to give out such information.

14. As far as the city administration is concerned, Firefighters should follow orders from company officers because it will:

A. Decrease the hazards of the job for him/her and peers.
B. Establish a good example for his/her peers.
C. Lessen dissension in the Fire Department.
D. Minimize confusion in emergency situations.

15. Of the choices below, which illustrates the best reason for Firefighters wearing of uniforms is:

A. To furnish clothing appropriate for the Firefighters job.
B. To increase the morale of Firefighters.
C. To illustrate the semi-military aspect of Fire Departments.
D. To identify Firefighters from other citizens in the community

16. The main benefit of prompt, thorough-complete reports is:

A. Accelerate official matters.
B. Provide a good source of reference documents.
C. Influence chief officers.
D. Depict the conscientious report writers.

17. Of the two block & tackles below, which one will lift a load more easily?

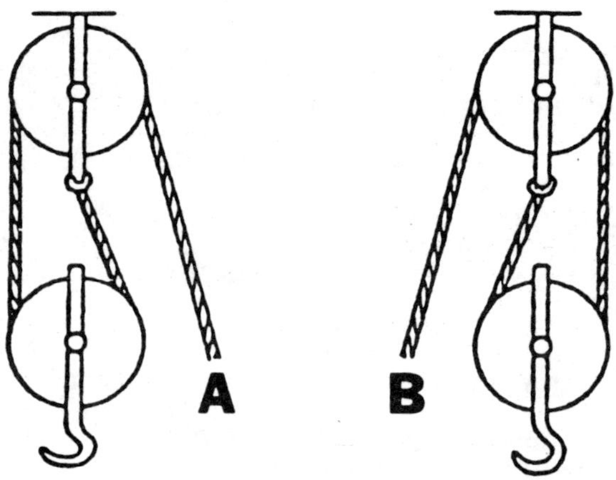

A. Drawing #1 B. Drawing #2 C. No difference.

18. In the drawing below:

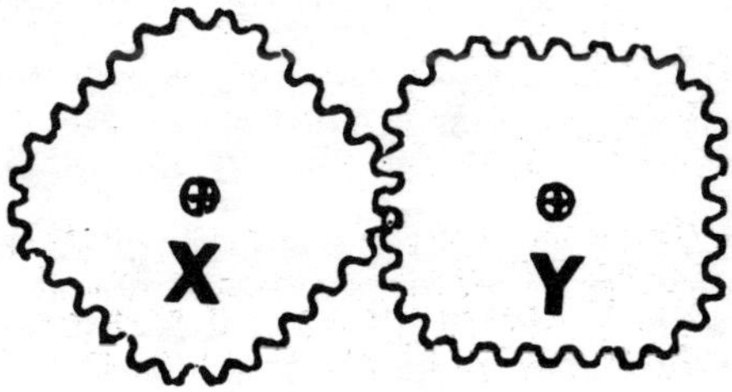

A. For each turn of gear "X", gear "Y" will make a complete rotation.
B. Only gear "X" can be the driving gear.
C. Wheel "Y" will move more slowly than Wheel "X".
D. None of the above.

19. In the drawing below, which ladder is the LEAST likely to tilt over?

A. Ladder #1 B. Ladder #2 C. makes no difference.

20. The object shown below is a:

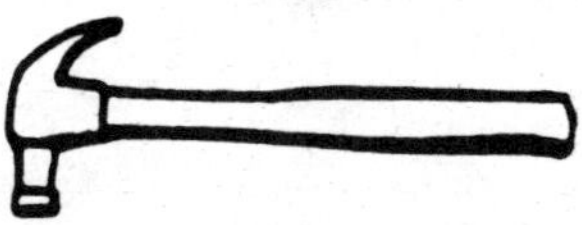

A. Ballpeen hammer.
C. Claw hammer.
B. Framers hammer.
D. Tack hammer

21. The object shown below is used to:

A. Start engines.
B. Replace spark plugs.
C. Drill holes.
D. Change auto tires.

22. The object shown below is used to:

A. Measure depth.
B. Measure thickness.
C. Punch holes.
D. Test metal hardness.

23. Two weights of the same size but of unequal weights are dropped at the same time from the same height:

A. The lightest weight will hit the ground first, because will be greater on the heavier weight.
B. The heaviest weight will hit the ground first, because it weighs more.
C. The two weights will hit the ground at the same time, because effect of gravity is the same on both weights.
D. The two weights will hit the ground at the same time, because they are the same size.

24. When drilling through wood, you should clamp an additional piece of wood on the underside, to:

A. Guide the drill bit.
B. Drill through faster.
C. Stabilize the drill bit.
D. Prevent the wood from splintering.

25. A tension pulley properly used on a belt connecting two pulleys of different size:

A. Will commonly be a crowned tension pulley.
B. Is generally placed nearer the large pulley.
C. Is generally placed nearer the small pulley.
D. In no way adds to the belt life.

26. When a ladder truck is traveling on level ground, stops suddenly, the crew will pitch forward because of:

A. Poor brakes.
B. Gravity.
C. Weightlessness.
D. Inertia.

27. Of the choices below, which one would least affect the needle of a compass:

A. Glass jar.
B. Iron bucket.
C. Another magnet.
D. Steel tools.

28. When the temperature gauge of a fire pumper indicates that the engine is overheating:

A. Add cool water without delay.
B. Add hot water without delay.
C. Add a cooling anti-freeze without delay.
D. Allow engine to cool down.

29. Prior to the adoption of modern construction codes, the factor that usually contributes the most to the spread of fire within multi-story buildings is:

A. Inferior alarm systems.
B. Inadequate water supply.
C. Unprotected vertical openings.
D. Unsafe building materials.

30. Of the choices below, which is the term that is used for a vertical section of wall that is one masonry unit in thickness:

A. Crown.
B. Wythe.
C. Connection.
D. Interval.

31. The commonly used figure for the density (weight divided by volume) of fresh water is:

A. 62.1 pounds per cubic foot.
B. 65.0 pounds per cubic foot.
C. 62.4 pounds per cubic foot.
D. 64.1 pounds per cubic foot.

32. Sea water is ______ compared to fresh water:

A. The same density.
B. Less dense.
C. More dense.
D. Sometimes more dense and sometimes less dense.

33. What is the mechanism that allows one rear wheel to go faster than the other, when a fire apparatus is going around a corner?

A. Drive shaft.
B. Rear axle.
C. Differential.
D. Universal joint.

34. Of the choices below, which is the term used when the end of a metal bar is held in the flame of a torch and the other end becomes too hot to handle?

A. Conduction.
B. Convection.
C. Ignition.
D. Radiation.

35. A volatile liquid is a liquid that will:

A. Heat slowly.
B. Resist oxidation.
C. Vaporize easily.
D. Vaporize slowly.

36. Which of the following statements is correct:

A. Water heated at sea level will boil at a higher temperature than water heated on the top of a mountain.
B. A large quantity of water will boil at a higher temperature than a small quantity.
C. Water heated slowly by a low flame will boil at a higher temperature than water heated quickly by a high flame.
D. Water always boils at the same temperature regardless of pressure.

37. Earthenware jugs will keep water cool, even in hot temperatures, because:

A. Particles of earthenware dissolve in the water and lower the waters temperature.
B. The rough surface of the jug radiates heat more rapidly.
C. The change of some of the water to a vapor lowers the temperature.
D. The jug is a good conductor of heat.

38. The threads on metal pipe that is used for liquids and gases will always be:

A. Straight.
B. Fine.
C. Coarse.
D. Tapered.

39. The following principle that applies to a jet plane is:

A. Energy can be neither created or destroyed.
B. If pressure is applied to a confined gas, the volume of the gas will decrease.
C. For every action there is an equal and opposite reaction.
D. Every effect has a cause.

40. The chain and the sprocket of a bicycle increase the:

A. Power to the rider.
B. force applied to the rear wheel.
C. Force applied to the road.
D. Speed of the rear wheel.

41. When a Firefighter enters a smoke filled structure, he/she will usually find the best viability near the:

A. Doors and windows.
B. Floor.
C. Center.
D. Ceiling.

42. Of the choices below, what would be the result of drying rubber lined hose at a very high temperature?

A. The hose would dry more thoroughly.
B. Hose would be too hot to handle.
C. The rubber lining would deteriorate too fast.
D. The cotton jacket would dry faster than the rubber lining.

43. Of the choices below, which would be the best explanation for a ship to appear to go below the horizon when you look out over the ocean from the beach:

A. The refraction of the sun.
B. The high and/or low tides.
C. Gravity.
D. The earths curvature.

44. Of the choices below, which one best explains the reason for there to be less of a difference of temperatures, from summer to winter, in the coastal areas as compared to the inland areas

A. The water in the coastal areas absorbs the summer heat and then releases it in the winter.
B. The prevailing winds change direction from summer to winter.
C. The warm water currents help to keep the coastal areas warm in the winter.
D. The sand in the coastal areas absorbs the summer heat and then releases it in the winter.

45. Of the choices below, which best describes the action taken when the gears of a fire apparatus are shifted from "high" gear to "low" gear:

A. The driving wheel will increase in speed.
B. The power in the engine will be increased.
C. The force from the driving wheel will be increased.
D. The force from the driving wheel will be decreased.

COMPLETE THE FOLLOWING ALPHABETICAL AND NUMERICAL PROGRESSIONS: (QUESTIONS #46 - #55)

46. T R P N __

A. M B. O C. L D. Q

47. C E H L __

A. M B. O C. L D. Q

48. AA DD GG JJ ___

A. KK B. LL C. MM D. OO

49. 20 19 21 19 22 19 23 ___

A. 19 B. 20 C. 21 D. 22

50. 0 1 3 6 10 15 21 ___

A. 22 B. 24 C. 28 D. 30

51. 2 6 18 54 162 ___

A. 164 B. 168 C. 216 D. 486

52. 96 24 28 30 10 13 14 ___

A. 7 B. 17 C. 28 D. 96

53. 2 2 3 3 5 5 8 ___

A. 7 B. 8 C. 9 D. 10

54. 11 2 12 3 13 4 14 ___

A. 1 B. 2 C. 5 D. 7

55. 4 8 16 8 16 32 24 ___

A. 38 B. 48 C. 58 D. 68

56. Firefighters don't have the time to study emergency incident manuals during life and death situations.

ACCORDING TO THE ABOVE STATEMENT

A. Property is not of the same value as a human life.
B. Firefighters don't have the time to study emergency manuals
C. Firefighters don't benefit from reading emergency manuals
D. During an emergency incident is not the time to prepare for an emergency incident.

57. Select from the choices below, the word that most nearly means **WARRANT**:

A. Justify B. Reserve C.Investigate D. Conclude.

58. Select from the choices below, the word that most nearly means **CONSTRAIN**:

A. Restrict B. Accumulate C. Reside D. Accuse

59. Select from the choices below, the word that most nearly means **DISCERN**:

A. Concede.
B. Understand.
C. Confirm.
D. Delay.

STATEMENT

At one time it was common practice for Firefighters to use large amounts of water to extinguish small fires, and in many instances the contents, of the structure involved, were not damaged by the fire but received damage from the large amounts of water used to extinguish the fire.

ACCORDING TO THE PREVIOUS STATEMENT

60.

A. Water must be located near the fire scene.
B. Large amounts of water are needed to extinguish all fires.
C. Fire losses are not always the result of fire.
D. Water without the use of nozzles is almost useless.

STATEMENT

The sprinkler system that is operated automatically or manually by a quick opening hydraulic or mechanical valve, and is a system of open sprinkler heads, or a combination of open and closed sprinkler heads is referred to as a deluge sprinkler system.

ACCORDING TO THE PREVIOUS STATEMENT

61. From the choices below, select the one that best describes an operation of a deluge sprinkler system from the preceding statement.

A. Deluge sprinkler systems use very little water.
B. Deluge sprinkler systems are operated on the principle of hydraulic pressure.
C. Deluge sprinkler systems may be controlled by a hydraulic valve which can be operated by hand.
D. Deluge systems that are hydraulically controlled are more effective than systems that are mechanically controlled.

STATEMENT

The fire prevention and fire protection questions resolve themselves into averting onset of fire, thwarting the threatening extension of fire, and affording quick discovery and extinguishment of fire.

ACCORDING TO THE PREVIOUS STATEMENT

62.

A. Most jurisdictions have a serious problem in fire detection.
B. In most jurisdictions, fire detection is a factor in fire prevention.
C. Fire prevention is significantly more important than fire detection.
D. Fire extinguishment is more important than fire prevention.

STATEMENT

To safeguard against overheating of electrical conductors in structures, a protective mechanism is supplied with each circuit. The device is designed to open the circuit and shut the flow of current off whenever the current surpasses a predetermined limit. The fuse, which is the most common form of "over-current" protection, contains a fusible metal element which melts when the current reaches a specified temperature, thus opening the circuit and stopping the flow of current.

ACCORDING TO THE PREVIOUS STATEMENT
(questions #63 & #64)

63. If a circuit is not carrying an electric current, it is:

A. An open circuit.
B. A closed circuit.
C. A circuit protected by a fuse.
D. A circuit protected by an over-current device.

64. The maximum number of circuits that can be handled by a fuse box containing 12 fuses is:

A. 6 circuits.
B. 12 circuits.
C. 24 circuits.
D. None of the above.

65. Of the following, which is the best example of a conductor:

A. A fire apparatus battery.
B. The metal fender of a vehicle in contact with a source of electricity.
C. An electrical wire carrying an electrical current.
D. An alternator.

STATEMENT

Unlined linen fire hose is basically a fabric tube manufactured of tightly woven linen yarn. Because of the inherent aspects of linen, soon after water is inserted into the fire hose the weave swells so that the tiny gaps between them shut, causing the hose to be almost watertight. This type of fire hose is inclined to deteriorate very fast if not thoroughly dried following use or if located where it will be exposed to moisture or the weather. It is not normally constructed to handle constant use or to be utilized where the fabric will be exposed to chafing from rough or sharp edges.

ACCORDING TO THE PREVIOUS STATEMENT
(questions #66 -#68)

66. Water leaks from unlined linen fire hose when water is first introduced would indicates that the fire hose is:

A. Not defective if it only leaks on the surface.
B. Defective.
C. Not defective if the leak is relative to the pressure introduced to the fire hose.
D. Not defective if the leak is substantially decreased when the fire hose becomes completely soaked.

67. Unlined linen fire hose is best utilized as:

A. Fire hose on fire apparatus.
B. A garden hose.
C. Fire hose at the Fire Academy.
D. Fire hose for fire emergencies in structures.

68. Of the choices below, the one that would be the least suitable location for the use of unlined linen fire hose, is:

A. Hotel.
B. Lumber yard.
C. Office building.
D. Manufacturing plant.

STATEMENT

The fact that some buildings are old is enough to give them a unique extent of fire risk. These old buildings will have out-of-date electrical wiring systems and installation of newer heating equipment for which the structures were not designed may augment to an the elevated danger. Old buildings have often been remodeled; portions of the structure may pre-date building codes; unsafe defects may cover them up. Generally speaking, older structures usually comprise of larger amounts of wood than similar newer structures which, by itself, makes the older structures a greater fire risk. The moisture content of the lumber in older buildings may or may not be less than that of newer buildings, that depends upon the environment that the older structures are exposed to.

ACCORDING TO THE PREVIOUS STATEMENT
(questions #69 - #72)

69. The main reason that old buildings present a unique fire hazard is:

A. Incompetent implementation of building codes.
B. Inadequate design of building when it was built.
C. Inconsistent modification of old buildings.
D. Failure to replace old equipment.

70. It is implied that lumber:

A. Is a greater fire risk than newer types of construction materials.
B. Should not be used in the construction of buildings.
C. Should not be used near electrical equipment.
D. Will lose moisture at a fixed rate.

71. The measure of moisture in the wooden areas of a structure depends upon the:

A. How old the building is.
B. Climate in the area of the structure.
C. Classification of heating equipment used in the structure.
D. Quality of wood utilized.

72. As far as building codes are concerned:

A. There are buildings that were built prior to such codes.
B. Building codes do not apply to old buildings.
C. Building codes are inadequate.
D. Electrical wires are not covered by building codes.

73. Select from the choices below, the word that most nearly means **ZEAL**:

A. Enthusiastic.
B. Integrity.
C. Generosity.
D. Trust.

74. Select from the choices below, the word that most nearly means **CONSUME**:

A. Stretch.
B. Cleanse.
C. Absorb.
D. take.

75. From the choices below, select the word that most nearly means **ETHICAL**:

A. Conscious.
B. Moral.
C. Courageous.
D. Auspicious.

CHOOSE THE **CORRECT** SPELLING FROM THE FOLLOWING CHOICES: (questions #76 - #86)

76.
A. abdoeman.
B. abdomen.
C. abdoman.
D. none of the above.

77.
A. acelerator.
B. accelorator.
C. accelerator.
D. none of the above.

78.
A. accordion.
B. acordyian.
C. accordyon.
D. none of the above.

79.
A. battalion.
B. batallion.
C. batalloun.
D. none of the above.

80.
A. bimetalic.
B. bimetallic.
C. bymetallic.
D. none of the above

81.
A. bowlyn.
B. bowline.
C. bowlynn.
D. none of the above.

82.
A. cardopulmonary.
B. cardyopulmonary.
C. cardiopulmonary.
D. none of the above.

83.
A. dalmation.
B. dalmatian.
C. dalmatain.
D. none of the above.

84.
A. deluge.
B. delouge.
C. daluge
D. none of the above.

85.
A. endurence.
B. endurince.
C. endurance.
D. none of the above.

86.
A. expander.
B. expandor.
C. expandar.
D. none of the above.

FOR QUESTIONS #87 - #90, SELECT THE SENTENCE THAT CONTAINS AN ERROR IN GRAMMAR, USAGE OR PUNCTUATION:

87.
A. The oral board was unanimous in its selection.
B. Members of the engine crew were continuously arguing with each other.
C. Two-fifths of this building is under water.
D. Three fifths of structures have been damaged.

88.
A. After the food has been cooked, it taste sourly.
B. There are times when pain is good.
C. The firefighter was hired at noon.
D. He earns a large income, but he spends his money senselessly.

89.
A. Ignorantly the firefighter performed the orders of the fire chief, allegedly a crazy man.
B. A patient, who is lying outside bleeding, needs our immediate attention.
C. In many ways the firefighter is intelligent; however, intelligence is not everything.
D. It is the consensus that the effort should be made.

90.
A. In all your days, did you ever observe such timidity?
B. Before you really see the object, you will have advanced well beyond it.
C. Did you take a boat ride in that ice-cold, glacier-fed Lake?
D. The firefighter was shocked by the watchdog.

CHOOSE THE CORRECT ANSWER FOR THE FOLLOWING MATH PROBLEMS: (QUESTIONS #91 - #98)

91. Add:

1/3 + 1 1/3 =

A. 2/3 B. 1 C. 1 1/2 D. 1 2/3 E. None

92. Subtract:

1/3 - 1/4 =

A. 1/8 B. 1/6 C. 1/10 D. 1/12 E. None

93. Multiply:

1/8 X 1/4 =

A. 1/8 B. 1/16 C. 1/24 D. 1/32 E. None

94. Divide:

3/4 by 2/3 =

A. 1 1/6 B. 1 1/3 C. 1 1/8 D. 1 E. None

95. 46 is what per cent of 200 ?

A. 23% B. 46% C. 92% D. 50% E. None

96. A strike team leaves for a brush fire at 2115 hours and arrives at the scene at 2445 hours. The strike team averaged 31 miles per hour. How many miles did they travel?

A. 105.8 miles.
B. 108.0 miles.
C. 108.5 miles.
D. 118.5 miles.

97. What will be the cost of 300 feet of fire hose at a cost of $1.457 per foot?

A. $437.10
B. $431.70
C. $423.10
D. $418.70

98. How much would Firefighter Murray pay for 5 "Pen-Lights" at a rate of $6.00 per dozen?

A. $1.50
B. $2.00
C. $2.50
D. $3.00

99. Of the choices below, which might be considered as a burn caused by radiated heat?

A. A Firefighters touches the window of a building on fire and burns his hand.
B. A Firefighter steps on a burning ember and burns his foot.
C. A vacationing Firefighter in Hawaii receives a severe sunburn on the beach.
D. A Firefighter is scalded by steam while fighting a fire using a fog stream.

100 Of the choices below, which one is the best reason that a Firefighter should know first aid?

A. To be able to give immediate relief to patients and prevent medical emergencies from escalating.
B. So that Firefighters will treat patients more compassionately.
C. To keep medical bills down for the citizens of the jurisdiction
D. To convince the citizens of the jurisdiction that the Firefighters have job dedication.

PRACTICE EXAM #7
ANSWERS

1. C
2. A
3. D
4. B
5. C
6. A
7. D
8. D
9. C
10. A
11. B
12. A
13. D
14. D
15. D
16. A
17. C
18. A
19. A
20. C
21. D
22. A
23. C
24. D
25. C
26. D
27. A
28. D
29. C
30. B
31. A
32. C
33. C
34. A
35. C
36. A
37. C
38. D
39. C
40. D
41. B
42. C
43. D
44. A
45. C
46. C
47. D
48. C
49. A
50. C
51. D
52. A
53. B
54. C
55. B
56. D
57. A
58. A
59. B
60. C
61. C
62. B
63. A
64. B
65. C
66. D
67. D
68. B
69. C
70. A
71. B
72. A
73. A
74. C
75. B
76. B
77. C
78. A
79. A
80. B
81. B
82. C
83. B
84. A
85. C
86. A
87. D
88. A
89. B
90. C
91. D
92. D
93. D
94. C
95. A
96. C
97. A
98. C
99. C
100. A

PRACTICE EXAM

#8

FIREFIGHTER WRITTEN

PRACTICE EXAM #8

1. Electrical currents are composed of moving particles called:

   A. protons
   B. neutrons
   C. electrons
   D. atoms
   E. molecules

2. When heated, nearly all materials:

   A. gain weight
   B. lose energy
   C. shrink
   D. expand
   E. melt

3. "Heat" and "Temperature"are:

   A. alike as both indicate a quantity measurement
   B. alike as both indicate a measurement of intensity
   C. different in that "heat" is a measure of intensity while "temperature" is a quantity measurement
   D. different in that "heat" is quantity measurement and "temperature" is a measure of intensity

4. The one of the following fire gases which causes an increase in breathing rate is:

   A. carbon monoxide
   B. carbon dioxide
   C. sulfur dioxide
   D. nitrogen dioxide

5. A rotating body tends to move away from the point around which it revolves. The direction of motion is:

   A. along the radius
   B. tangential
   C. centripetal
   D. circular

6. The viscosity of an oil is:

   A. its tendency to coagulate when heated
   B. the measure of its resistance to flow
   C. its rate of atomization
   D. the speed with which it flows through pipe lines

7. If boiling water is poured into a drinking glass, the glass is likely to crack. If, however, a metal spoon is first placed in the glass, it is much less likely to crack. The reason that the glass with the spoon is less likely to crack is that the spoon:

A. distributes the water over a larger surface of the glass
B. quickly absorbs heat from the water
C. reinforces the glass
D. reduces the amount of water which can be poured into the glass

8. It is considered poor practice to increase the leverage of a wrench by placing a pipe over the handle of the wrench:

A. because the wrench may break
B. because the wrench may slip from the nut
C. because it is harder to place the wrench on the nut
D. because the wrench is more difficult to handle

9. In the basement of a hotel there is a 20 foot section of an 8 inch metal pipe. The pipe is supported by three metal straps from the basement ceiling. If each end of the pipe is supported by a strap and the center strap is situated closer to one end than the other, which of the following is CORRECT?

A. the end strap that is the farthest away from the center strap holds the greatest weight
B. the end strap closest to the center strap holds the greates weight
C. the center strap holds the greatest weight
D. all three straps will hold an equal weight

10. In the drawing below:

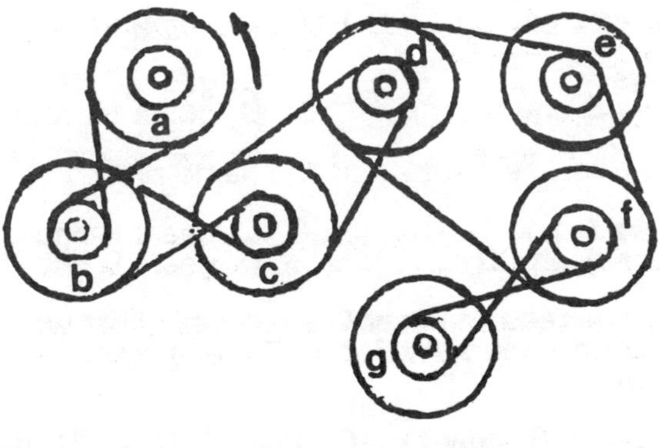

A. wheels A and G turn in the same direction
B. wheel B and E turn in opposite directions
C. wheel C makes more revolutions than wheel G
D. wheel B makes more revolutions than wheel E

11. In the drawing below, by crossing the pulley belt:

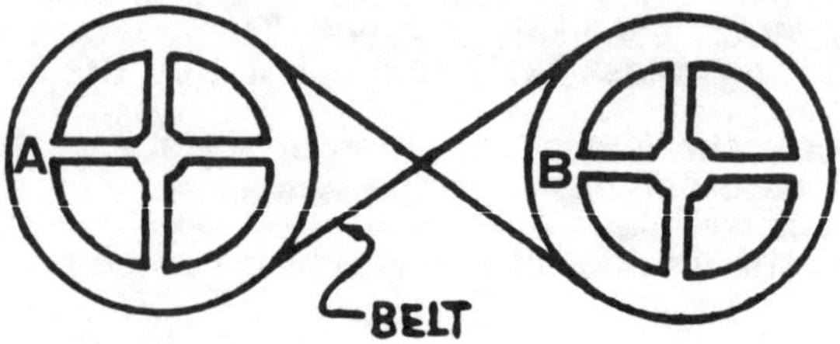

A. the slack will be taken up
B. the wheels will turn in opposite directions
C. the belt will last longer
D. wheel A will turn faster than wheel B

QUESTIONS 12, 13, 14 AND 15 ALL RELATE TO THE DRAWING BELOW WITH GEAR A, THE DRIVING GEAR, ROTATING IN THE DIRECTION OF THE ARROW. GEARS A AND D HAVE TWICE AS MANY TEETH AS GEAR B. GEAR C HAS FOUR TIMES AS MANY TEETH AS GEAR B.

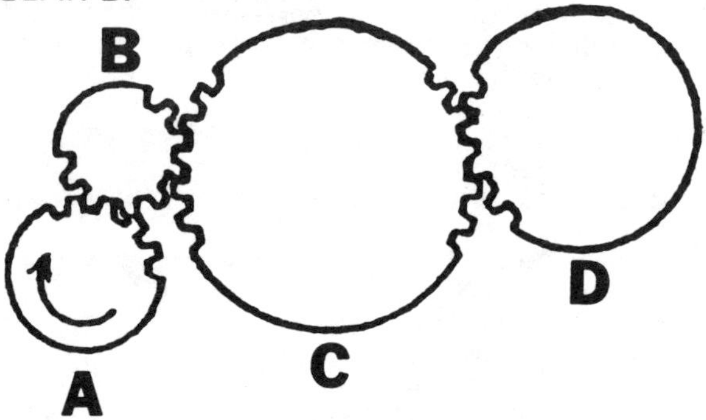

12. Which two gears turn in the same direction?

A. gears A and B
B. gears B and C
C. gears C and D
D. gears B and D

13. Which two gears rotate at the same rate?

A. gears A and C
B. gears A and D
C. gears B and C
D. gears B and D

14. If all of the teeth on gear C are knocked off without influencing the teeth on gears A, B, and D, then rotation would only take place with:

A. gear C B. gear D C. gears A & B D. gears A, B, & D

15. If gear D rotates at the rate of 100 rpm then gear B rotates at a rate of:

A. 10 rpm B. 25 rpm C. 50 rpm D. none of these

16. Carbon monoxide gas is best described as:

A. pink colored
B. having a strong odor
C. tasting like bitter almonds
D. slightly lighter than air

17. Generally speaking, the smoke in a burning building will continue to build up until:

A. the fire goes out
B. water is put on the fire
C. all the fuel is used up
D. the building is ventilated

CHOOSE THE NUMBER THAT COMES NEXT IN THE FOLLOWING NUMERICAL PROGRESSIONS:

18.
1 2 6 7 21 22 66 __

A. 19 B. 23 C. 67 D 68

19.
4 8 9 13 14 18 19 __

A. 15 B. 17 C. 23 D. 25

20.
25 27 30 15 17 20 10 __

A. 9 B. 12 C. 18 D. 22

21.
6 8 10 8 10 12 10 __

A. 4 B. 12 C. 14 D 16

CHOOSE THE LETTER THAT COMES NEXT IN THE FOLLOWING ALPHABETICAL PROGRESSIONS:

22.
C E H L __

A. O B. P C. Q D. R

23.
AA DD GG JJ __

A. KK B. LL C. MM D. NN

IN THE FOLLOWING DRAWINGS, CHOOSE AN ARTICLE FROM THE RIGHT THAT MAY BE MADE FROM THE FLAT DESIGN AT THE LEFT

24.

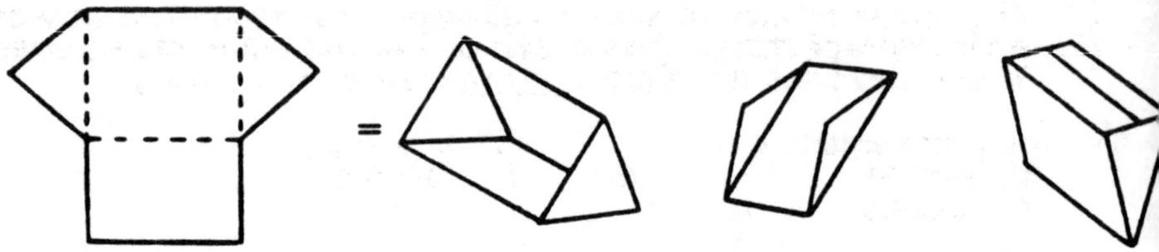

DESIGN A. B. C.

25.

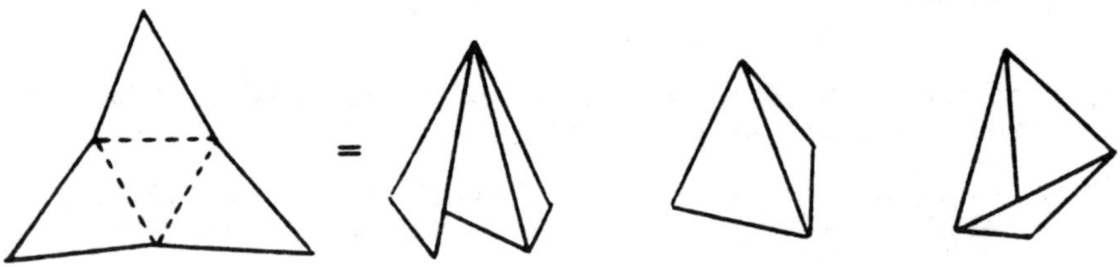

DESIGN A. B. C.

26. Radiated heat finds difficulty in travelling through water or air heavily laden with moisture, and it is by impregnating the atmosphere with water that the radiant heat given off by a fire can be absorbed. According to this statement, it would be most accurate to assert that radiant heat is best absorbed at a fire by:

A. charging the air with water spray
B. filtering the atmosphere with moisture
C. filtering the atmosphere with carbon dioxide
D. making the air impregnable

27. To say that a fire is incipient means most nearly that the fire is:

A. innocuous B. beginning C. intentional D. viable

28. An abrasive was required by the machinist to complete his task. As used in the sentence, the word abrasive means a substance used for:

A. coating
B. lubricating
C. measuring
D. polishing

29. The facades of the building were dirty and grimy." As used in this sentence, the word facades means most nearly:

A. cellars
B. fronts
C. residents
D. surroundings

30. The stores refusal to accept delivery of the merchandise was a violation of the express provisions of the contract. As used in this sentence the word express means most nearly:

A. transparent
B. implied
C. penalty
D. disputed
E. elaborate

31. A statement of self evident truth is called a(n):

A. aphorism
B. axiom
C. hypothesis
D. theory

32. The wire connecting the two terminals must be kept taut. As used in this sentence, the word taut means most nearly:

A. without defects
B. without slack
C. without electrical charge
D. without pressure

READ THE FOLLOWING STATEMENT AND THEN ANSWER QUESTIONS 33 - 37 ACCORDING TO WHAT YOU READ IN THE STATEMENT.

STATEMENT

Fire often travels inside the partitions of a burning building. Many partitions contain wood studs that support the partitions. The studs leave a space for the fire to travel along. Flames may spread from the bottom to the upper floors through the partitions. Sparks from a fire in the upper part of a partition may fall and start a fire at the bottom. Some signs that a fire is spreading inside a partition are:

1. blistering paint
2. discolored paint or wallpaper, or
3. partitions that feel hot to the touch.

If any of these signs is present, the partition must be opened up to look for the fire. Finding cobwebs inside the partition is one sign that fire has not spread through the partition.

ACCORDING TO THE PREVIOUS STATEMENT

33. Fires can spread inside partitions because:

A. there are spaces between studs inside of partitions
B. fires can burn anywhere
C. partitions are made out of materials that burn easily
D. partitions are usually painted or wallpapered

34. The main point of the passage is:

A how fires spread inside partitions
B. how cobwebs help firefighters
C. how partitions are built
D. how to keep fires from spreading

35. If a firefighter finds that the paint on a partition is beginning to blister, he should first:

A. wet down the partition
B. check the partitions in other rooms
C. chop a hole in the partition
D. close windows and doors and leave the room

36. Cobwebs inside a partition are a sign that fire has not spread inside the partition because:

A. cobwebs are fire resistant
B. fire destroys cobwebs easily
C. spiders do not build cobwebs near fires
D. cobwebs fill up the spaces between studs

37. One way to tell if fire is spreading inside a partition is the:

A. temperature of the partition
B. color of the smoke
C. age of the plaster
D. spacing of the studs

38. Exertion means most nearly:

A. argument
B. deferral
C. induce
D. effort
E. planning

39. Gigantic means most nearly:

A. elaborate
B. up-to-date
C. automated
D. huge
E. permanent

40. Permanent means most nearly:

A. renowned
B. continual
C. common
D. artistic
E. well-planned

41. Reliable means most nearly:

A. dependable
B. easy to use
C. productive
D. accomplished
E. fast-acting

42. Indicate means most nearly:

A. relate
B. keep secret
C. point out
D. retrace
E. walk along

43. Positive means most nearly:

A. sure
B. dependent
C. neutral
D. contradictory
E. reliable

44. Launch means most nearly:

A. conclude
B. start
C. pursue
D. navigate
E. wake

45. Redundant means most nearly:

A. foolish
B. appalling
C. aromatic
D. excessive
E.reclining

READ THE FOLLOWING STATEMENT AND THEN ANSWER QUESTIONS 46 - 50 ACCORDING TO WHAT YOU READ IN THE STATEMENT.

STATEMENT

Unless they have had a fire, most people are not aware of the things firefighters do. Too often the public think of firefighters as lounging around the firehouse between fires. Firefighters can help change this image in small ways by their appearance, by greeting visitors who come to the firehouse, by their behavior on the street at a fire, and by treating the public in a courteous manner. For example, 90% of the rescues made by the average fire department take place at relatively small fires, not at spectacular extra alarm fires. The public rarely hears about them because the fire departments seldom let the press know about firefighters who have performed acts of bravery at routine fires.

ACCORDING TO THE PREVIOUS STATEMENT

46. What are firefighters doing when not fighting fires?

A. lounging around the firehouse
B. working on public relations projects
C. making repairs to the equipment
D. the passage doesn't say

47. What would be the best title for this passage?

A. An Inside Look at the Fire Department
B. Making the Most of Fire Prevention Week
C. Improving the Fire Departments Public Image
D. Brave Acts Performed by Firefighters

48. Most of the rescue made by firefighters take place at:

A. extra alarm fires
B. special emergencies where no fire is involved
C. relatively small fires
D. spectacularly large fires

49. The public rarely hears about rescues made by firefighters at routine fires because:

A. information about fires must be kept confidential
B. Fire departments seldom report these rescues to the press
C. most of these rescues take place late at night
D. reporters are not interested in covering routine fires

50. The passage places responsibility for improving the fire departments image on the:

A. fire department itself
B. press
C. public.
D. people rescued by firefighters

51. Hot gases cause a room to burn by:

A. creating a draft
B. exploding
C. giving off sparks
D. raising the room temperature

52. One millimeter equals what portion of a meter?

A. one-thousandth of a meter
B. one-tenth of a meter
C. one half of a meter
D. one meter

53. Fire Officers should know that the substance that would require the least thickness to lessen atomic radiation by one-half would be:

A. water B. earth C. concrete D. lead

54. Fog fire streams, will divided the water into a great number of small droplets. Skillful firefighters should therefore understand that a fog fire stream is more capable of absorbing heat than solid fire stream of water since a fog fire-stream:

A. has a greater surface area
B. has a greater density
C. is more continuous
D. dissolves lass air

55. As a fire increases in temperature, the percentage of carbon monoxide in the gases liberated:

A. remains constant
B. increases
C. decreases
D. decreases at a fast rate

56. It is important that firefighters, in cold climates, realize that at a specific temperature a wet hand will freeze to exposed metal, but not to a piece of wood because:

A. the metal expands and contracts more than the wood
B. the wood is more pliable than the metal
C. the wood will ignite at a lower temperature than the metal
D. the metal is a better conductor of heat than the wood

57. When metal pipe is used to carry liquids or gases, the threaded portion is always:

A. straight B. tapered C. fine D. coarse

58. Hydraulics is the branch of science which deals with:

A. water or other fluid in motion
B. water under high pressure
C. methods of firefighting
D. gravity of water
E. centrifugal force

CHOOSE THE CORRECT SPELLING OF THE WORD IN THE FOLLOWING: (questions #59 & #60)

59.
A. carpenter
B. carponder
C. carponter
D. none of the above

60.
A. hydrolic
B. hydralic
C. hydraullic
D. none of the above

CHOOSE THE MISSPELLED WORD IN THE FOLLOWING: (questions #61 - #65)

61.
A. lieutenant
B. preservation
C. anaesthetic
D. questionaire

62.
A. obedient
B. vengeance
C. plague
D. none misspelled

63.
A. aisle
B. proposal
C. people's
D. condemed

64.
A. atheletic
B. queue (waiting line)
C. schedule
D. furl

65.
A. negate
B. chagrined
C. associate
D. none misspelled

ONE SENTENCE IN EACH OF THE FOLLOWING QUESTIONS CONTAINS AN ERROR IN GRAMMAR, USAGE, OR PUNCTUATION. CHOOSE THE SENTENCE WHICH CONTAINS THE ERROR.
(questions #66 - #68)

66.
A. everyone went excluding Bert and me.
B. It influences you no more than us.
C. I like those kind of nozzles best.
D. Everybody, except my two sisters, is going to the station.
E. Lay the nozzles on the shelf.

67.
A. The Jones'es children came in uniforms.
B. Keats's poems are read with enjoyment.
C. Hardy's novels are still popular.
D. The fault, I am afraid, is yours.
E. He attributed the pain to the food he had eaten.

68.
A. Can you, both of you, afford to miss it?
B. I talked with Jack and Jill.
C. I talked with both Jack and Jill.
D. He is wiser than us all.
E. Who shall I say telephoned?

69. If these words : fire, rang, the, times, bell, three, were arranged to make a sentence, the third word would be:

A. fire
B. rang
C. times
D. bell
E. the

70. Add:

.55 + .34 =

A. .86 B. .87 C. .89 D. .90

71. Subtract:

3,742 - 1,096 =

A. 2,658 B. 2,636 C. 2,646 D. 2,746

72. Divide:

12,462 by 62 =

A. 200 B. 201 C. 211 D. 190

73. Find the percentage: 75% of 6,400 =

A. 1,600 B. 4,800 C. 3,200 D. 5,200

74. Multiply:

19.261 X 25 =

A. 481.525 B. 481.516 C. 4.81516 D. 481.515

75. Divide:

9 into 6.444 =

A. 716 B. .716 C. 71.6 D. .715

76. The County Fire Department found that it had prevented major damage in 80% of the fires reported. It had prevented such damage in 560 calls. The total number of fires reported was:

A. 600 B. 660 C. 700 D. 750

77. A custom canvas salvage cover measures 6 feet by 9 feet. The largest circular area that can be covered completely by this salvage cover has a diameter of:

A. 9 ft B. 8 ft C. 7 ft D. 6 ft

78. One quarter of a million is the same as:

A. 2,500,000
B. 25,000
C. 250,000
D. 2,500

79. How many more ounces of wheel bearing grease are contained in a 10-ounce container than a half-pound container?

A. 6 ounces B. 5 ounces C. 22 ounces D. 2 ounces

80. Firefighter Wendl borrowed $4000.00 at 14% interest. How much additional money can he borrow at 16% if his total interest is not to exceed $1250.00 a year?

A. $800.00 B. $900.00 C. $2000.00 D. $4000.00

81. If twelve gallons of fuel drove one ladder truck a distance of 188.4 miles and the same amount of fuel took another fire apparatus a distance of 202.8 miles, how much better mileage per gallon has the second fire apparatus than the first:

A. 1.7 MPG
B. 5 MPG
C. 12 MPG
D. 14.4 MPG.
E. none of the above.

82. A tool dealer has put a "firefighters tool kit" on sale with discounts of 25% and 10% from $60, the marked price. How much will it cost a firefighter to buy the tool kit?

A. $13.50
B. $21.00
C. $39
D. $40.50
E. None of the above

83. The following fire gases are a danger to human life. Which is the one that tops the list in the cause of fire deaths:

A. nitrogen
B. carbon monoxide
C. hydrogen
D. carbon dioxide

STATEMENT

Building owners expect to see fire streams start as soon as firefighters arrive on scene and generally do not have patience when firefighters take time to develop heavy streams during the early stages of a fire when they can do the most good.

ACCORDING TO THE ABOVE STATEMENT

84.

A. Building owners are generally well informed on the best fire fighting techniques
B. firefighters should get fire streams started on a fire immediately even though the streams are small in order to satisfy the building owner
C. heavy streams during the early stages of a fire are seldom necessary
D. firefighters frequently waste time working to get heavy streams started when the building owner knows that many smaller streams would do the most good
E. the building owner may not understand the importance of taking the extra time required to get large streams going early

85. Of the following which is the best reason for fire departments to have ladders of different sizes:

A. fires develop at different heights
B. firefighters are more active than most citizens
C. firefighters are taller than most citizens
D. structures are not fireproof

86. All of the following are the names of time zones except:

A. Central
B. Mountain
C. gulf
D. Pacific

87. The Police Department normally will establish "Fire lines" in order to keep spectators out of the immediate vicinity of a fire incident while firefighters are performing their duties. Of the following, the best justification for the establishment of these fire lines is to:

A. prevent theft
B. prevent interference of operations performed by firefighters
C. allow privacy to the victims of the fire
D. help capture of arsonists in the crowd

88. As a construction material, the physical toughness of wood is more inconstant than steel or concrete. Consequently:

A. it is used solely in the construction of floors
B. it should never be utilized in conjunction with steel or concrete
C. it requires higher safety considerations in design
D. the use of wood is substantially decreasing

89. Listed below are five operating characteristics of most automatic sprinkler systems. The one characteristic of those listed which is LEAST desirable is that automatic sprinkler systems:

A. function only in the fire zone
B. go into application soon after a fire starts
C. open in the midst of high heat and smoke
D. continue operating after the fire is extinguished

90. Fire Officers should recognize that if the water draining away from the fire is cold, it may be assumed that:

A. the water has performed a insignificant role in putting out the fire
B. the hoses are positioned to the windward
C. additional fire hose lines are needed to extinguish the fire
D. the hydrant pressure is too low

STATEMENT

One or more fire apparatus may be called to a location by sending on the signaling device therein the following signals in the order given: the preliminary signal 777, the number of the street box, and the number to indicate how many apparatus are needed. In accordance with these instructions, the proper signal for calling four fire apparatus to Box 423 is:

ACCORDING TO THE PREVIOUS STATEMENT

91.

A. 4-777-423
B. 777-423-4
C. 7777-423
D. 7-423-4

92. During a high rise fire incident a firefighter is ordered, by the captain to go to the third floor to fight the fire which is breaking through there. At the foot of the third floor stairs the firefighter stumbles over a victim who has apparently been overcome by smoke. Which one of the following should the firefighter do first:

A. carry the victim out immediately
B. complete the assignment and then come back for the victim
C. run to the window and shout for help
D. administer CPR to the victim

93. Firefighters should be aware that spalling of concrete is caused by:

A. not enough cement in the concrete mix
B. extremely high temperatures
C. exposure to the weather
D. chemical damage

94. Of the choices below, which is the best reason for firefighters need to wear helmets:

A. gives firefighters a military appearance
B. differentiates the firefighters from the public
C. provides the firefighters protection from water and falling debris
D. allows company officers the ability to identify firefighters by companies

95. When firefighters uncoil rope in the wrong direction, it will causes it to twist and kink. New rope should be uncoiled:

A. from the outside of the coil in a counterclockwise direction
B. from the center of the coil in a counterclockwise direction
C. from the center of the coil in a clockwise direction
D. from the outside of the coil in a clockwise direction

96. Fire Officers should be aware that in case of a fire in a U.S. mail box it is recommend that an extinguishing agent which smothers the fire should be used. Of the following, the most likely reason for NOT recommending the use of water is that:

A. water is not effective on fires in small, tightly enclosed spaces
B. someone might have mailed chemicals that could explode in contact with water
C. Water may damage the mail untouched by fire so that it could not be delivered
D. the smothering agent can be put on the fire faster than water can be

97. Which of the following perform their duties under the supervision of a physician?

A. paramedics
B. first responders
C. emergency medical technicians
D. none of the above

98. Firefighting is generally considered a function of:

A. a local government
B. the federal government
C. a fire protection district
D. a contract fire department

99. Fire Officers should be aware that when responding to a fire incident, that they should not to talk to the apparatus driver except to give orders or directions. Of the following, the best reason for this rule is that:

A. it gives the officer an opportunity to make preliminary plans for handling the fire problem
B. enables the apparatus driver to concentrate on driving the apparatus
C. maintains the proper relationship between the ranks while on duty
D. permits the officer to observe the apparatus driver skill, or lack of skill, in driving the apparatus

100 It is important for firefighters to have some knowledge of first aid, the best reason for this would be to:

A. make the public aware that firefighters care about their health
B. protect the public from doctor bills
C. make firefighters more compassionate and understanding
D. give unhealthy and injured patients immediate assistance

PRACTICE EXAM #8
ANSWERS

1. C	26. A	51. D	76. C
2. D	27. B	52. A	77. D
3. D	28. D	53. D	78. C
4. B	29. B	54. A	79. D
5. B	30. D	55. B	80. D
6. B	31. B	56. D	81. E
7. B	32. B	57. B	82. D
8. A	33. A	58. A	83. B
9. A	34. A	59. A	84. E
10. B	35. C	60. D	85. A
11. B	36. B	61. D	86. C
12. D	37. A	62. D	87. B
13. B	38. D	63. D	88. C
14. C	39. D	64. A	89. D
15. D	40. B	65. D	90. A
16. D	41. A	66. C	91. B
17. D	42. C	67. A	92. A
18. C	43. A	68. D	93. B
19. C	44. B	69. D	94. C
20. B	45. D	70. C	95. B
21. B	46. D	71. C	96. C
22. C	47. C	72. B	97. A
23. C	48. C	73. B	98. A
24. A	49. B	74. A	99. B
25. B	50. A	75. B	100. D

PRACTICE EXAM

#9

FIREFIGHTER WRITTEN
PRACTICE EXAM #9

1. BARGAIN means most nearly:

   A. agreement B. debt C. routine D. design

2. ADJOURN means most nearly:

   A. start B. handle C. attend D. complete

3. Engine #81 is shorter than Aerial Ladder #81 but longer than Rescue #81. Hose Wagon #81 is shorter than Aerial Ladder #81, but because of its length, cannot fit into Fire Station #2. Engine #81 will just fit in Fire Station #2. Which one of the following is the longest piece of fire apparatus that will fit in Fire Station #2:

   A. Hose Wagon #81
   B. Aerial Ladder #81
   C. Rescue #81
   D. Engine #81

STATEMENT

Arsonists are people who set fires deliberately. They don't look like criminals, but they cost the nation millions of dollars in property loss, and sometimes loss of life. Arsonists set fires for many different reasons. Sometimes a shop keeper sees no way out of losing his business, and sets fire to it so he can collect the insurance. Another type of arsonist wants revenge, and sets fire to the home or shop of someone he feels has treated him unfairly. Some arsonists just like the excitement of seeing the fire burn and watching the firefighters at work; arsonists of this type have been known to help fight the fire.

ACCORDING TO THE PREVIOUS STATEMENT
(questions #4 - #7)

4. An arsonist is a person who:

   A. intentionally sets a fire
   B. likes watching fires
   C. wants to retaliate
   D. lacks money

5. Arsonists have been known to assist in fighting fires because they:

   A. feet ashamed
   B. relish the excitement
   C. desire to gain money
   D. do not wish to injure anybody

6. Business owners sometimes become arsonists in order to:

   A. commit suicide
   B. collect insurance money
   C. camouflage a crime
   D. increase their prices

7. The primary point of this statement is that arsonists:

   A. would make reliable firefighters
   B. are not criminals
   C. are mentally ill
   D. are not all alike

STATEMENT

If there are quickly available avenues of escape from fire and proper planning for their prompt use in case of emergency, the danger of loss of life is largely eliminated, no matter how severe the fire hazard may be.

ACCORDING TO THE PREVIOUS STATEMENT

8.

   A. fires seldom happen in movie theaters
   B. suitable fire exits are a vital factor in the prevention of deaths in a fire
   C. the fire hazard is voided when proper fire escapes are provided
   D. all emergencies need strategy

STATEMENT

Fire pumping apparatus in the fire service are used to take water from the hydrants at whatever pressure is available and boost that pressure to whatever is necessary for water streams. The pressure at hydrant level with flowing water is lower than the pressure for a closed hydrant. Pressures necessary for water streams are measured when the required water flow is being delivered. The minimum satisfactory water pressure for firefighting purposes is 20 lbs at the flowing hydrant. This amount is required to give an adequate pressure for the pumper supply hose.

ACCORDING TO THE PREVIOUS STATEMENT

9.

   A. 20 lbs of pressure is the lowest capacity for the fire pumping apparatus
   B. the fire pumping apparatus are mainly used to measure water pressures
   C. if the pressure at a closed hydrant is 20 lbs, the pressure when flowing from the hydrant will be greater
   D. fire pumping apparatus need a minimum of 20 lbs pressure from the discharging hydrant

STATEMENT

Cotton fabrics treated with the "Flame Defiant Process" have features which cause them to be far better than any previously known flame-retardant-treated cotton fabrics. "Flame Defiant Process" treated fabrics are durable to repeated laundering and dry cleaning; are glow resistant as well as flame resistant; when exposed to flame or intense heat form tough, pliable, and protective chars; are inert physiologically to persons handling or exposed to the fabric; are only slightly heavier than untreated cotton fabrics; and are susceptible to further wet and dry finishing treatments. In addition, the treated fabrics exhibit little or no adverse change in feel, texture, and appearance, and are shrink, rot, and mildew-resistant. The treatment decreases strength only slightly. Finished cotton fabrics have "easy care" properties in that they are wrinkle resistant and dry rapidly.

ACCORDING TO THE PREVIOUS STATEMENT
(questions #10 - #14)

10. It is accurate to say that the writer of the above statement presents:

A. details but attains no resolution concerning the value of the process
B. his conclusion concerning the value of the process and facts to support this conclusion
C. his resolution concerning the value of the process unsupported by facts
D. neither fact nor conclusions, but simply describes the process

11. The one of the following articles for which the "Flame Defiant Process" would be most suitable is:

A. nylon hosiery
B. woolen shirt
C. silk shirt
D. cotton shirt

12. The one of the following features of the "Flame Defiant Process" which is not addressed in the above statement is its effects on:

A. costs
B. washability
C. wearability
D. the human body

13. The principal justification for treating a fabric with the "Flame Defiant Process" is to:

A. make ready the fabric for other wet and dry finishing treatment
B. make it shrink, rot, and mildew-resistant
C. enlarge its weight and strength
D. decrease its chance of catching fire

14. Of the following, which is the one that would be regarded as minor disadvantage of the "Flame Defiant Process" is that it:

A. develops chars when exposed to fire
B. causes fabrics to become mildew-resistant
C. adds to the weight of fabrics
D. is congruous with other finishing treatments

15. Continuing without PLANNING is as ill-advised to firefighting as in any other endeavor. PLANNING means most nearly:

A. harmony B. surety C. accuracy D. preparation

16. FORTITUDE means most nearly:

A. affluence
B. determination
C. forthright
D. thunderous
E. destiny

17. DISCREDIT means most nearly:

A. apprise
B. discourage
C. debate
D. dishonor
E. rescue

18. BONA FIDE means most nearly:

A. light hearted
B. amenable
C. authentic
D. benevolent
E. fashionable

19. ACUTE means most nearly:

A. critical
B. acrid
C. invigorating
D. valid
E. assured

20. RELUCTANT means most nearly:

A. impotent
B. reliant
C. unwilling
D. continue
E. remorseful

IN THE FOLLOWING EXAMPLES CHOOSE THE SENTENCE THAT IS THE MOST CORRECT AS FAR AS GRAMMAR, USAGE, AND PUNCTUATION: (questions #21 - #24)

21.

A. When a quality actress cries, she feels real sad.
B. When a quality actress cries, she feels real sadly.
C. When a quality actress cries, she feels really sadly.
D. When a quality actress cries, she really feels sad.
E. When a quality actress cries, she feels really sad.

22.

A. Bill wails at Cindy that is her, not he, who shouts.
B. Bill wails at Cindy that is her, not him who shouts.
C. Bill wails at Cindy that it is she, not him who shouts.
D. Bill wails at Cindy that it is she, not he, who shouts.
E. Bill wails at Cindy that it is her, not he who shouts.

23.

A. Ted and Bruce are more poised than he, but Bill is the brighter of all three firefighters.
B. Ted and Bruce are more poised than he, but Bill is the brightest of all three firefighters.
C. Ted and Bruce are more poised than him, but Bill is the brightest of all three firefighters.
D. Ted and Bruce are more poised than him, but Bill is the brighter of all three firefighters.
E. Ted and Bruce are most poised than him, but Bill is the brightest of all three firefighters.

24.

A. The Fire Chief's speech opened with the following statement:"The intention of the Fire Department is, to make the people aware of one of the greatest dangers to the safety of the country."
B. The Fire Chief's speech opened with the following statement, "The intention of the Fire Department is to make the people aware of one of the greatest dangers to the safety of the country."
C. The Fire Chief's speech opened with the following statement:"The intention of the Fire Department is: to make the people aware of one of the greatest dangers to the safety of the country."
D. The Fire Chief's speech opened with the following statement:"The intention of the Fire Department is to make the people aware of one of the greatest dangers to the safety of the country."
E. The Fire Chief's speech opened with the following statement:"the intention of the government is to make the people aware of one of the greatest dangers to the safety of the country."

CHOOSE THE SENTENCE IN THE FOLLOWING QUESTION WHICH CONTAINS AN ERROR IN GRAMMAR, USAGE, OR PUNCTUATION.

25.
A. "I should not have reached this position," he said, "if the way had been made too smooth."
B. "Can you show us any reason," said the delegate, "Why the company cannot afford to grant this increase?"
C. I walked two miles farther than I had intended.
D. "If the driver of a passing car had not shouted 'Your wheel is loose!'" he said, "I shudder to think of what might have happened."
E. My hero in high school was the football and basketball coach.

26. Which of the following is most unlike the other four?

A. policeman
B. fireman
C. conductor
D. minister
E. postman

CHOOSE THE MISSPELLED WORD IN THE FOLLOWING QUESTIONS: (questions 27 -

27.
A. hopeless
B. absense
C. foretell
D. certain

28.
A. option
B. extradite
C. comparitive
D. jealousy

29.
A. together
B. attempt
C. loyality
D. innocent

30.
A. relieve
B. disappear
C. development
D. matress

31.
A. falsify
B. blight
C. flexible
D. drasticaly

32.
A. tenements
B. sulphuric
C. acetylene
D. turpentine

IN THE FOLLOWING, CHOOSE THE CORRECT SPELLING OF THE WORD SHOWN: (questions #33 & #34)

33.
A. fundamental
B. fundemental
C. fundimental
D. none of these

34.
A. chimmney
B. chimney
C. chimny
D. none of these

35. If you as a firefighter were to enter a house in which you found all the occupants overcome by gas, the first thing you should do is:

A. call for immediate assistance
B. summon the inhalator squad
C. open all windows
D. apply artificial respiration

36. The architects of today build skyward mainly for:

A. safety
B. economy
C. fresh air
D. light
E. beauty

37. Small arteries branch to form a network of capillaries. In turn, the capillaries unite to form:

A. alveoli
B. arteries
C. auricles
D. veins
E. ventricles

38. Fire Officers should be aware that of the following, the primary obstacle in the way of acquiring reliable information about these causes of fires is that:

A. firemen are too busy putting out fires to have time for investigation of the causes of fires
B. most people have little knowledge of fire hazards
C. fires destroy much of the evidence that would indicate the causes of the fires
D. fire departments are more interested in fire prevention than in investigating fires that have already occurred

39. Firefighters are cautioned to avoid wearing rings on their fingers. The principal justification for this advice is that the rings have a affinity to:

A. be damaged during firefighting operations
B. scrape victims receiving first aid treatment
C. catch on things and injure the firefighter
D. damage furniture and/or other valuable objects

40. Tapping on pipes is a good way to locate victims because:

A. firefighter can use Morse code
B. sound travels through a pipe
C. firefighters can signal each other this way
D. pipes usually are not covered by rubber

41. Most fire authorities agree that of the following tools the one that is the <u>least</u> essential in fire fighting:

A. axe
B. wire cutter
C. ladder
D. hammer

42. Normally, sprinkler heads have to be replaced each time they are used. The best reason as to why this is required is that the sprinkler heads:

A. are exposed to rusting after discharging water
B. may become obstructed after discharging water
C. have a deformed pattern of water discharge after use
D. are set off by the effect of heat on metal and cannot be reset

43. A fire company inspecting buildings in a commercial area and inspecting a structure with an exterior that appears to be of natural stone. The building owner tells the firefighters that it is not necessary to inspect his building as it is "fireproof". The firefighters, nevertheless, complete the inspection of the building. Of the following, the best justification for continuing the inspection is that:

A. stone buildings incur fire as easily as wooden buildings
B. the fire department should not make exceptions in its inspection procedures
C. the structure may be built of imitation stone
D. interiors and contents of stone buildings can catch fire

44. Company Officers should recognize that if after a fire has been extinguished, the fire company has been asked to go back to the third floor of the fire structure. Which of the following acts, the one which Company Officers should take first is to:

A. determine if it is safe to go onto the third floor
B. investigate for any fire extension that may remain on the third floor
C. position something over holes which have been cut in the floor
D. withdraw any broken timbers or glass found on the third floor

45. Fire authorities agree that from a fire protection point of view, the most desirable type of construction for exterior walls is:

A. masonry veneer
B. masonry
C. metal
D. wood frame

46. Fire Officers should be aware that when the fire apparatus is ready to depart the fire station, to a fire incident, prior to all firefighters getting completely dressed and equipped, sometimes to avoid delay, these firefighters may finish dressing on the way to the fire. This practice is undesirable mainly because:

A. a substandard impression is made on the public
B. the firefighters are not able to size-up the conditions as they approach the fire.
C. the chance of losing equipment from the moving apparatus is increased
D. the danger of injury to the firefighters is increased

47. A school principal made a practice of holding fire drills on the first Monday of each month, just prior recess. In general, conducting fire drills according to a regular schedule is:

A. good; students will be more cooperative when fire drills result in early recess
B. poor; fire drills should not be anticipated
C. good; panic is averted if the students know that there isn't a "real" fire
D. poor; holding fire drills once or twice a term is sufficient

48. The U.S. Fire Administration and the National Fire Academy are located in:

A. Boston, MA
B. Emmitsburg, MD
C. Stillwater, OK
D. Washington D.C.

49. Firefighters should recognize that the basic geometric figure from which trusses are composed are called:

A. trapezoids
B. tetrahedrons
C. squares
D. triangles

50. A fire Captain orders a firefighter to open the windows in a room filled with smoke. The firefighter begins with the window nearest the entrance and follows the wall around the room until all the windows are open. The most important reason for using this procedure is that he can:

A. avoid tripping over furniture
B. breathe the purest air near the walls
C. avoid the weakened floor in the middle of the room
D. find his way back to the entrance

51. Gothic architecture is known for its:

A. brick walls
B. pointed arches
C. round domes
D. low ceilings

52. A fire pumping apparatus that is discharging 300 lbs. per square inch pressure, delivers water through 100 feet of fire hose laid horizontally. If the nozzle at the end of the hose is shut so that no water can flow, then the pressure at the valve is, for practical purposes:

A. greater than the pressure at the fire pumping apparatus
B. equal to the pressure at the fire pumping apparatus
C. less than the pressure at the fire pumping apparatus
D. greater or less than the pressure at the pump, depending on the type of fire pumping apparatus

53. In the drawing below:

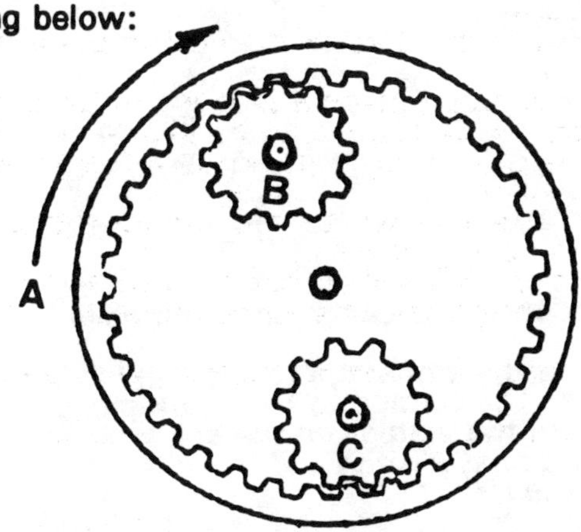

A. Gear A and gear B will turn in opposite directions
B. Gear B and gear C will turn in opposite directions
C. answers A and B are both correct
D. none of the above

54. In the drawing below, when a force is exerted at the location of the #1 on the teeterboard, in which direction will the #2 go?

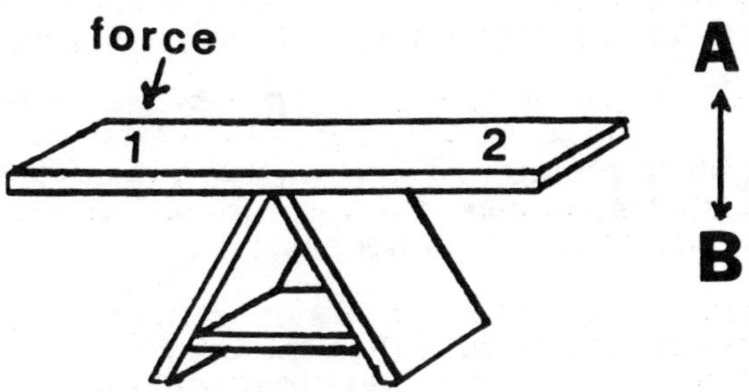

A. up
B. down
C. neither direction

55. Of the two drawings below, which winch will be able to lift a heavier load?

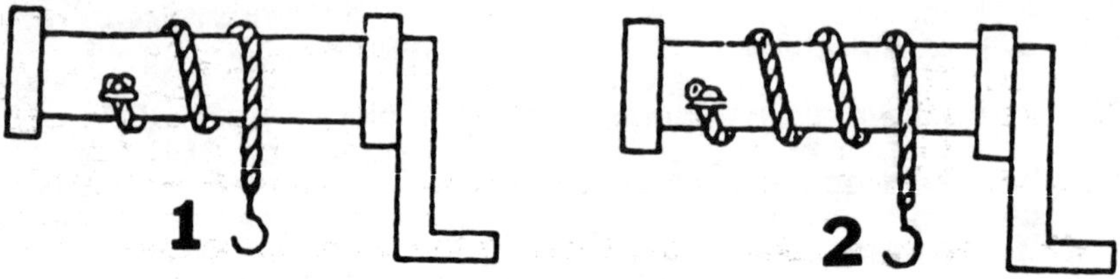

A. Drawing #1 B. Drawing #2 C. equal loads

56. In the drawing below, the maximum weight that can be lifted with a 75 lbs pull as shown is:

A. 50 lbs
B. 75 lbs
C. 100 lbs
D. 150 lbs

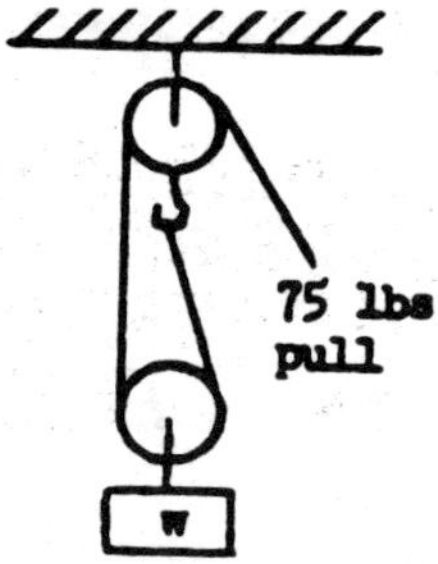

57. The function of the coil in an automobile is to:

A. increase the amperage to the spark plugs
B. increase the voltage to the spark plugs
C. decrease the voltage to the spark plugs
D. act as a relay switch and close the circuit
E. break the circuit

58. Firefighters should realize that fire departments have various sizes of ladders primarily because some:

A. firefighters are larger than others
B. firefighters are stronger than others
C. fires are more difficult to extinguish than others
D. fires are located at greater heights than others

59. Mercury is how many times heavier than water?

A. 5 B. 9 C. 13 D. 16

60. Firefighters performing operations in cold climates should be aware that at a given temperature a wet hand will freeze to metal, but not to wood because:

A. metal enlarges and shrinks more than wood
B. wood is more pliable than the metal
C. wood will burn at a lower temperature than the metal
D. metal is a better conductor of heat than the wood

61. How many blocks are there in the drawing below?

A. 15
B. 16
C. 17
D. 18

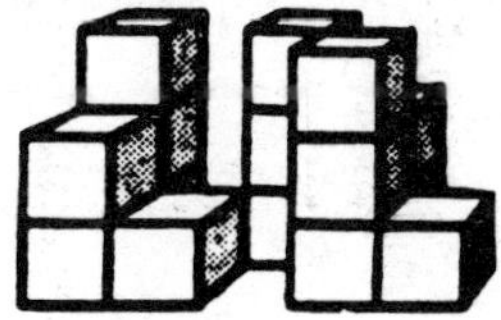

62. How many blocks are there in the drawing below?

A. 9
B. 10
C. 12
D. 13

IN THE FOLLOWING DRAWINGS DETERMINE THE RELATIONSHIP BETWEEN THE FIGURES LOCATED IN THE TOP TWO BOXES AT THE LEFT, THEN DETERMINE HOW THE FIGURE IN THE BOTTOM LEFT BOX IS SOMEHOW RELATED TO ONE OF THE FOUR FIGURES TO THE RIGHT. REMEMBER POSITION & DIRECTION

QUESTIONS 63 -65

63.

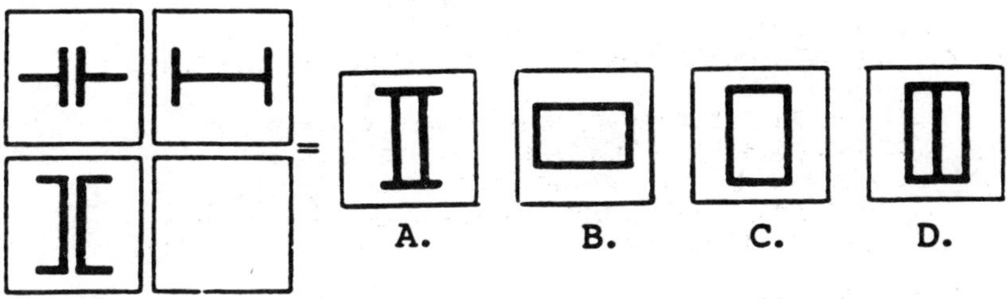

64.

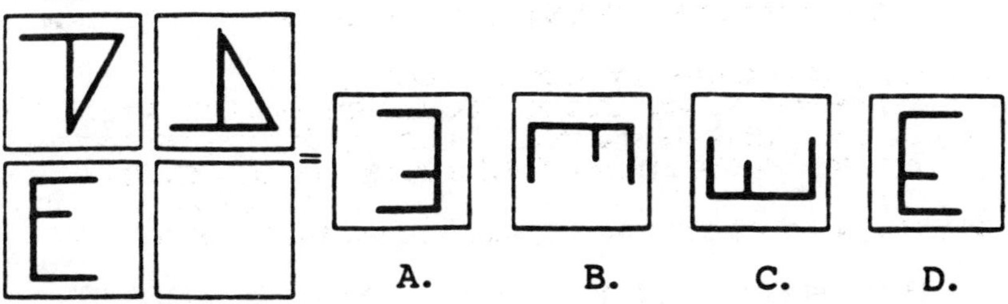

65.

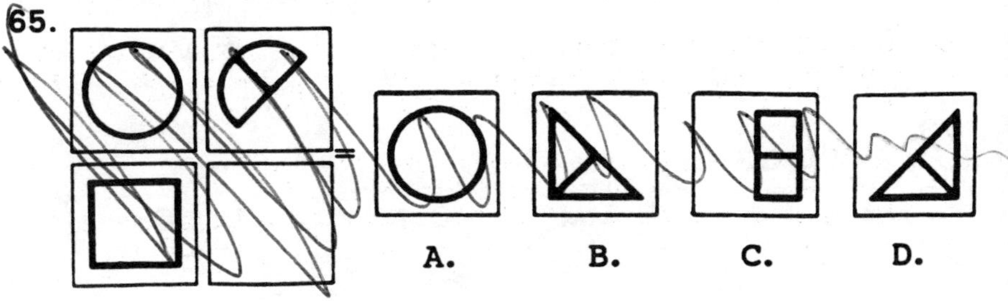

WHICH NUMBER COMES **NEXT** IN THE FOLLOWING NUMERICAL PROGRESSIONS: (questions #66 - #68)

66.
12 15 27 30 42 45 __

A. 47 B. 50 C. 52 D. 57

67.
4 6 12 14 28 30 60 62 __

A. 62 B. 98 C. 114 D. 124

68.
30 23 17 12 8 5 __

A. 2 B. 3 C. 1 D. 0

FIND THE LETTER THAT COMES **NEXT** IN THE FOLLOWING ALPHABETICAL PROGRESSIONS: (questions #69 - #73)

69.
B E J Q __

A. W B. X C. Y D. Z

70.
D F E H G __

A. I B. J C. K D. M

71.
C E H L __

A. Q B. R C. S D. T

72.
D F E H G __

A. J B. K C. L D. M

73.
AA DD GG JJ __

A. LL B. MM C. NN D. OO

74. Firefighters should be aware that of the following, the one that is a product of combustion is:

A. gasoline B. oxygen C. fire gases D. electricity

75. The term "flame" is best described as:

A. a noticeable product of incomplete combustion
B. a visible luminous body of burning gas
C. the vaporized result of combustion
D. energy measured in degrees of temperature

76. Add:

88.8 + 99.9 =

A. 178.8 B. 187.8 C. 188.7 D. 188

77. Divide:

7 into 336 =

A. 42 B. 46 C. 48 D. 49

78. Multiply

.7 X 5 =

A. .35 B. 35 C. 3.5 D. 350

79. Add:

755 + 284 + 709 =

A. 1,748 B. 1,757 C. 1,758 D. 1,848

80. Subtract:

833 - 654 =

A. 179 B. 188 C. 221 D. 279

81. Multiply:

86 X 4 =

A. 324 B. 326 C. 334 D. none of these

82. A special salvage cover measures 6 feet by 9 feet. The largest circular area that can be covered completely by this salvage cover has a diameter of:

A. 9 feet B. 8 feet C. 7 feet D. 6 feet

83. If water weighs 62 1/2 pounds per cubic foot, water one foot deep covering an area that is 10 feet by 16 feet will weigh most nearly:

A. 1,000 pounds
B. 6,000 pounds
C 10,000 pounds
D. 16,000 pounds

84. If a "tag line" 100 feet long is cut so that one piece is 2/3 as long as the other piece, the length of the longer piece must be:

A. 60 feet B. 66 2/3 feet C. 70 feet D. 75 feet

85. Round off 97,850 to the nearest thousand:

A. 97,000 B. 98,000 C. 100,000 D. 90,000

86. Paramedic Roberts can re-supply Rescue 82 in 25 minutes; Paramedic Cota can re-supply the rescue in 50 minutes, but Paramedic Cannon can re-supply the rescue in 10 minutes. How long would it take all three Paramedics together to re-supply the rescue?

A. 10 min B. 8 1/3 min C. 6 1/4 min D. 5 7/11 min

87. If 8x represents the perimeter of a rectangle and 2x + 3 represents its length, what is its width?

A. 10x + 3 B. 2x - 3 C. 3 D. 6x - 3

88. A square container 18 feet on a side has the same area as a rectangular container. If one side of the rectangular container is 9 feet, what is the number of feet in the other dimensions:

A. 9 B. 2 C. 27 D. 36

STATEMENT

The backdraft commonly referred to in Fire Service is an explosion caused by the ignition of mixtures of air with minute particles of unburned carbon and gaseous matter in smoke, resulting from imperfect combustion of organic substances. Such explosions frequently occur in burning buildings where, due to insufficient air, complete combustion was prevented and to which a sudden supply of air has been furnished by opening a door or window.

ACCORDING TO THE PREVIOUS STATEMENT

89.

A. smoke in burning structures is normally highly combustible
B. lack of adequate air may render a combination of smoke and air potentially hazardous
C. abrupt opening of a door or window in a burning structure will induce an explosion
D. backdrafts typically develop in structures that are well entilated

STATEMENT

To extinguish a large fire entails the choice of hose streams of a size suitable to the amount of material burning or jeopardized and the vantage points from which they may be performed. Nothing is accomplished when a stream is directed randomly into smoke where no fire shows. Successful extinguishment includes getting in front of the fire and stopping its progress. It is frequently dictated to close in on the fire from all feasible directions, moving hose lines up quickly, as soon as the fire in any one place has been extinguished.

ACCORDING TO THE PREVIOUS STATEMENT

90.

A. the control of the extension of a fire is of rudimentary relevance in firefighting
B. when the fire cannot be seen it is best to discharge a heavy stream of water where the smoke is the heaviest
C. in fighting fire it is consistently best to approach it from the rear
D. the most influential consideration in fighting a fire is the choice of the fire streams
E. in general, once hose lines are laid they are rarely moved unless the fire threatens to extened

91. Which one of the following is **NOT** one of the ways that heat can travel through a burning building:

A. conduction
B. convection
C. radiation
D. conclusion

92. The fuel in the cylinder of a diesel engine is ignited by:

A. an electric spark
B. a supercharger
C. the injector
D. the heat of compression

93. The air we exhale, compared to the air we inhale, contains:

A. less carbon dioxide
B. more oxygen
C. less nitrogen
D. more water vapor

94. The method by which the sun transmits heat to the earth is known as:

A. conduction
B. radiation
C. convection
D. circulation

95. An electron is a/an:

A. electronic nucleus
B. static charge
C. particle of negative electricity
D. hydrogen atom

96. Geiger counters will react to:

A. both alpha and beta but not gamma radiation
B. both alpha and gamma but not beta radiation
C. both beta and gamma but not alpha radiation
D. alpha, beta, and gamma radiation

97. Radioactive fallout can be sensed by:

A. a dusty appearance in the atmosphere
B. a peculiar odor
C. a whirring noise
D. none of these

98. Which of thge following metals would be considered the LEAST adequate as a non-sparking metal for tools:

A. bronze
B. brass
C. copper alloys
D. hardened copper

99. Of the following metals, the one which is the best conductor of both heat and electricity is:

A. silver
B. copper
C. lead
D. nickel

100 The proper sequence for firefighters to search an area is:

A. from center to edge
B. from edge to center
C. around the edges
D. from corner to corner

PRACTICE EXAM #9
ANSWERS

1. A
2. D
3. D
4. A
5. B
6. B
7. D
8. B
9. D
10. B
11. D
12. A
13. D
14. C
15. D
16. B
17. D
18. C
19. A
20. C
21. D
22. D
23. B
24. D
25. D
26. D
27. B
28. C
29. C
30. D
31. D
32. B
33. A
34. B
35. C
36. B
37. D
38. C
39. C
40. B
41. D
42. D
43. B
44. A
45. B
46. D
47. B
48. B
49. D
50. D
51. B
52. B
53. D
54. A
55. C
56. D
57. A
58. D
59. C
60. D
61. A
62. C
63. C
64. D
65. A
66. D
67. D
68. B
69. D
70. C
71. A
72. B
73. B
74. C
75. B
76. C
77. C
78. C
79. A
80. A
81. D
82. D
83. C
84. A
85. B
86. C
87. B
88. D
89. B
90. A
91. D
92. D
93. D
94. B
95. C
96. C
97. D
98. B
99. A
100. B

PRACTICE EXAM
#10

FIREFIGHTER WRITTEN
PRACTICE EXAM #10

1. Select from the choices below, the word that most nearly means **ADMINISTRATION:**

   A. Management.
   B. Diplomatic.
   C. Assistant.
   D. Admiration.

2. Select from the choices below, the word that most nearly means **ACCRUE:**

   A. Account.
   B. Accumulate.
   C. Assimilate.
   D. Deferral.

3. Select from the choices below, the word that most nearly means **REMISS:**

   A. Ashamed.
   B. Sensitive.
   C. Negligent.
   D. Precise.

4. Preserving life from fire is the primary objective of fire prevention and fire protection. From the choices below, select the one that most nearly describes the meaning of the previous statement:

   A. The primary objective of any safety program is fire prevention and fire protection.
   B. Firefighters jeopardize their lives in the prevention of fires.
   C. Firefighter should be well trained in emergency medical techniques.
   D. In firefighting the protection of life is more important than the protection of property.

5. Select from the choices below, the word that most nearly means **INJURY:**

   A. Trauma.
   B. Misery.
   C. Danger.
   D. Demise.

6. Select from the choices below, the word that most nearly means **MERIT:**

   A. Promise.
   B. Order.
   C. Earn.
   D. Win.

7. More consistency has been realized in legislation for arson control with most jurisdictions than for any other fire prevention distinction. From the choices below, you could conclude that arson control:

A. Has been adopted by all jurisdictions.
B. Legislation is the same in all jurisdictions.
C. Is closely associated to fire prevention.
D. Is a immense obstacle for legislators.

STATEMENT

Fire strategy is the plan of attack on a fire. Fire strategy should make prime use of equipment and personnel and take into consideration fire behavior, the nature of the occupancy, environmental conditions, and weather factors. Fire tactics are the various maneuvers that can be employed in a strategy to successfully fight a fire. Even though no two fires are alike, it is possible to lay down general plans for firefighting operations primarily because the elements of similarity are sufficient enough to establish fire strategy and fire tactics that are applicable in a variety of situations. In any fire remember for covering all points of a fire, cover the front and rear, and over and under along with covering all exposures.

ACCORDING TO THE PREVIOUS STATEMENT
(questions #8 - #12)

8. Fire strategy is:

A. The attack on a fire.
B. The evaluation of a fire scene.
C. The plan of attack on a fire.
D. he critique of a fire.

9. Fire tactics are:

A. The extinguishment of fire.
B. The critique of a fire.
C. The various apparatus and equipment that may be employed in a strategy to successfully fight a fire.
D. The various maneuvers that can be employed in a strategy to successfully fight a fire.

10. Fire strategy should:

A. Make prime use of equipment and personnel
B. Take fire behavior into consideration.
C. Take the weather into consideration.
D. All of the above.

11. It is possible to lay down general plans for firefighting, because:

A. No two fires are alike.
B. All fires are alike.
C. The elements of similarity are sufficient enough to establish tactics and strategy.
D. The elements of tactics and strategy are always exactly the same.

12. In covering all points of a fire, officers should remember to cover:

A. The front and rear.
B. Over and under.
C. All exposures.
D. All of the above.

13. If a Firefighter were **ARROGANT**, which of the choices below describes how other Firefighters would feel about him/her:

A. Appropriate because the Firefighter would be self confident.
B. Inappropriate since the Firefighter would reduce the "espirt de corp".
C. Inappropriate since the Firefighter would not follow orders.
D. Appropriate because the Firefighter would set a good prototype for others.

STATEMENT

Flash point is the lowest temperature of a liquid at which it gives off vapor which forms an ignitible mixture with air at the surface of the liquid, or within the vessel it is in. This is the lowest temperature that a flame will occur with a heat source. Fire will not continue without a heat source. Flash point is the ignitible temperature of a liquids vapor. The flash point more than any other physical property determines the hazard of a liquid. The flash point is at a temperature of less than 5 degrees F. below fire point. Fire point is the temperature that a liquid is able to continue to burn without an outside heat source. Fire point is the lowest temperature of a liquid at which vapors are evolved fast enough to support continued combustion.

ACCORDING TO THE PREVIOUS STATEMENT
(questions #14 - #17)

14. Flash point of a liquid is:

A. The lowest temperature that a flame will occur, with a heat source.
B. The lowest temperature that a flame will occur, without a heat source.
C. The lowest temperature that a flame will continue without an outside heat source.
D. None of the above.

15. Fire point of a liquid is:

A. The lowest temperature that a flame will occur, with a heat source.
B. The lowest temperature that a flame will occur, without a heat source.
C. The lowest temperature that a flame will continue, without an outside heat source.
D. None of the above.

16. Flash point of a liquid is:

A. More than 5 degrees F. above fire point.
B. Less than 5 degrees F. below fire point.
C. The same temperature as fire point.
D. Plus or minus 5 degrees of fire point.

17. While conducting firefighting operations it is normally better to operate from the lee side of the fire whenever it is possible It can be deduced from the above statement that Firefighters normally perform firefighting operations:

A. On the side that the wind blows towards.
B. On the side that the wind blows away from.
C. On the side the right side of the fire whenever possible.
D. As far from the fire as feasible

18. **SLING** is related to **THROW** as **IMPEDE** is related to:

A. Detain.
B. Proceed.
C. Exhale.
D. Proclaim.

19. Cylinder #1 and cylinder #2 turn in opposite directions at the same time. If #1 turns in the direction of the arrow, in what direction will the pulley hook travel?

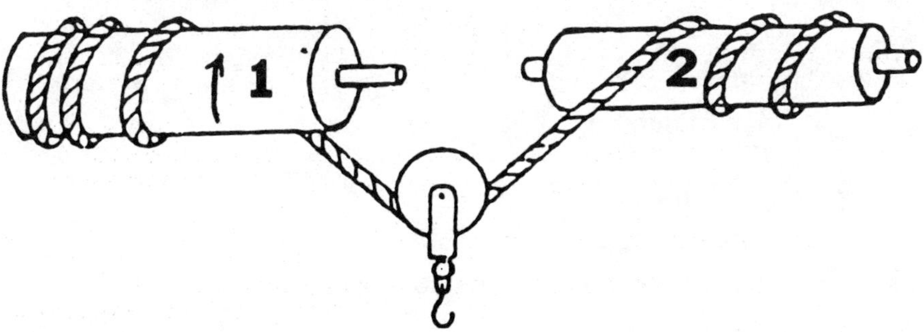

A. In an upward direction.
B. In a downward direction.
C. It will not travel up or down.

20. Which direction must the inside gear be rotated in order to rotate the outside gear?

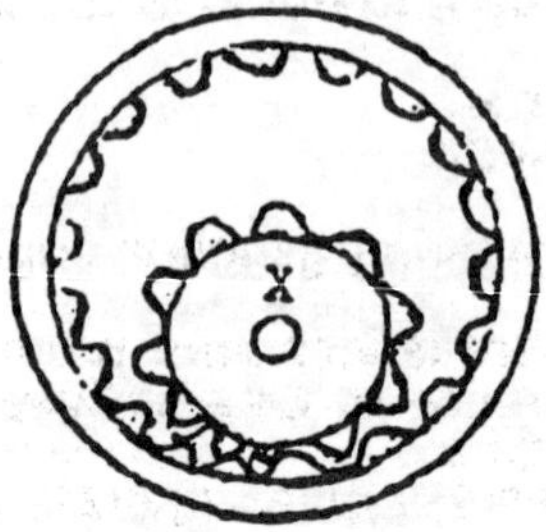

A. Counterclockwise.
B. Clockwise.
C. In either direction.

21. In the drawing below, which block will be the most difficult to tip over?

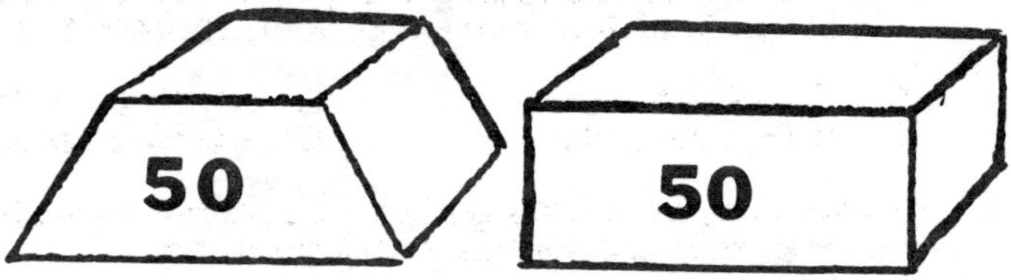

A. #1. B. #2. C. There would be no difference.

22. The object shown below is used to:

A. Tap pipes.
B. Turn nuts.
C. Turn philip screws.
D. Round corners of nuts.

23. The object shown below is a:

A. Transit knob.
B. Center punch.
C. Clock pendulum.
D. Plumb bob.

24. The object shown below is used:

A. For welding.
B. For soldering.
C. For lubricating.
D. For painting.

25. The use of an absorbent material placed on a slippery surface, so as to reduce the chances slipping, the affect of an absorbent material is to increase:

A. Force.
B. Energy.
C. Friction.
D. Gravity.

26. In comparing gasoline engines to diesel engines:

A. There is less soot produced by diesel engines.
B. The use of oil additives is not recommended in gasoline engines.
C. Diesel engines operate at higher temperatures.
D. Gasoline engines operate at higher temperatures.

27. A weight is to be supported from a brace by a chain of 30 links and a hook. Each link of the chain weighs 2 pounds and can support a weight of 1,000 pounds, and the hook weighs 10 pounds and can support a weight of 5,000 pounds, what is the maximum load that can be supported from the hook?

A. 25,000 pounds.
B. 5,000 pounds.
C. 1,000 pounds.
D. 930 pounds.

28. When a rod is submerged in a container of water it appears to be bent, this phenomenon is known as:

A. Diffraction.
B. Diffusion.
C. Refraction.
D. Reflection.

29. In the world of colors, red, yellow and blue are referred to as:

A. Primary colors.
B. Secondary colors.
C. Complementary colors.
D. Related colors.

30. A gallon of fresh water weighs:

A. 7.48 pounds.
B. 6.20 pounds.
C. 8.34 pounds.
D. 23.1 pounds.

31. The term used for the definition, the force exerted by water, neglecting any effects of motion or velocity, on a unit area is:

A. Static pressure.
B. Static head.
C. pressure head.
D. Choices A, B, and C are all correct terms.

32. Pressure head in a hydraulic system is measured in:

A. Pounds per square inch.
B. Pounds.
C. Inches of mercury.
D. Choices A, B, C, are all false.

33. Heavy timber construction is also known as:

A. Heavy frame construction.
B. High fire-resistive construction.
C. High fireproof construction.
D. Mill construction. (slow burning construction)

34. In many structures, the primary factor that allows the spread of fire is:

A. Stairs and shafts.
B. Unprotected ceilings and floors.
C. Unprotected windows.
D. Large continuous areas.

35. Of the choices below, which is the substance that will assume the state of gas, liquid or solid, by the change of temperature alone, is:

A. Air.
B. Water.
C. Carbon dioxide.
D. Hydrogen.

36. An ampere is a measure of:

A. Water flow.
B. Water pressure.
C. Electrical pressure.
D. Electrical flow.

37. In regards to physics, the term convection usually is associated with the:

A. Calculation of humidity.
B. Transfer of heat by moving masses of matter created by variation in density.
C. Most common method of securing specific gravity.
D. Transfer of heat by radiation.

38. Of the choices below, which one best describes the term "SPOT WELD":

A. A welding job where only one side is welded.
B. A welding job where the parts are welded together in various places to hold them in place.
C. A welding job that must be performed at a specific location.
D. A welding job that must use a heli-arc.

39. Firefighting ability, in general, is measured in terms of:

A. Ability to follow orders.
B. Physical strength.
C. Reasoning ability.
D. Intellectual superiority.

40. Of the following what is the best reason for children to be silent during school fire drills:

A. Children talking will annoy teachers.
B. Fire drill will take place with less confusion.
C. To show how well disciplined the children are.
D. To impress the children with discipline.

41. Of the following, what is the best reason for Firefighters to clean and check their apparatus and equipment after fires:

A. So they have will have more time for recreation.
B. So they will look busy.
C. To keep equipment and apparatus in good shape.
D. To keep apparatus and equipment looking good.

42. Of the following choices, which one is **NOT** a factor of the fire triangle:

A. Fuel.
B. Heat.
C. Oxygen.
D. Inhibited chain reaction.

43. The primary function of a Fire Department is to:

A. Promote good public relations.
B. Protect life and property.
C. Improve the city politics.
D. Cut down on industrial problems.

44. What should you do as a Firefighter, if while making a routine company fire inspection of a business, that has violations, the business owner becomes annoyed and states to you that he is a close and personal friend of the Fire Chief:

A. Politely continue the inspection.
B. Call for police assistance.
C. Get tough and write up every possible violation.
D. Find out how well the business owner knows the Fire Chief.

45. Of the following choices, which is the best reason for Firefighters attempting to confine a fire to its point of origin:

A. To minimize property damage.
B. To minimize fire risk.
C. To minimize length of hose lines.
D. To minimize amount of firefighting personnel.

46. Of the choices below, which type of business, if involved in a fire, would present the greatest poisonous gas hazard to Firefighters?

A. Paper mill.
B. Grocery store.
C. Drug store.
D. Furniture manufacturing.

47. An electric light bulb that is powered by **MORE** than its rated voltage will result in:

A. dimmer light with a shorter bulb life.
B. dimmer light with a longer bulb life.
C. brighter light with a shorter bulb life.
D. brighter light with a longer bulb life.

48. If you were spending the night in a hotel, and you were awakened by the smell of smoke, of the choices below the best thing to do when leaving the room is:

A. Feel the door before opening it.
B. Open the door as quickly as possible.
C. Open a window first then open the door.
D. Open the door and leave it open.

49. A Firefighter is cleaning the fire apparatus in front of the station when a citizen stops and advises him/her that some children are throwing large objects off a nearby overpass on to vehicles passing underneath. What action should the Firefighter take?

A. Advise the citizen that this is not a Fire Department issue.
B. Follow the citizen the overpass and stop the children from dropping the objects.
C. Take the citizen to the police station.
D. Report the incident to his/her superior officer.

50. On occasion, during a fire incident, Firefighters may suffer extensive punishment so they may attain a desired position which could have been reached faster and with less adversity. Of the following choices, the primary implication of the above statement is:

A. It is necessary for Firefighters to have endurance.
B. Training for Firefighters is an significant influence in firefighting.
C. Firefighters should stop during a fire incident and consider each obstacle existing.
D. Firefighters need to be alert for firefighting.

51. Suppose that, while fighting a fire, a Firefighter is on a ladder, by the window of a building. The Firefighter is ordered to break the window so that he/she can reach inside and open the lock on a nearby door. Which of the following would be the **LEAST** appropriate method of breaking the window glass?

A. Using a long handled tool such as an axe.
B. Opening the window pane closest to the door latch.
C. Taking position directly below the glass that will be broken.
D. Taking a position to the side of the window that is to be broken.

52. When fire hydrants are damaged by vehicles, it is usually because:

A. Drivers of the vehicles cannot see the fire hydrants.
B. The fire hydrants are usually unprotected.
C. Fire hydrants are usually located in the center of the block.
D. Apathy from the drivers, since the cost of the fire hydrants is very low.

53. If properly employed, water may be used effectively on flammable liquid fires. The previous statement would imply that water could be used on which type of fire listed below?

A. Phosphorus fires.
B. Petroleum fires.
C. Magnesium fires.
D. Lumber fires.

CHOOSE THE CORRECT ANSWER FOR THE FOLLOWING MATH PROBLEMS: (questions #54 - #65)

54. Add:

5/8 + 7/8 =

A. 1 B. 1 1/8 C. 1 1/2 D. 1 5/8 E. None

55. Subtract:

3/4 - 3/4 =

A. 1/4 B. 1/3 C. 1/8 D 1/10 E. None

56. Multiply:

.1 X .1 =

A. 1 B. .1 C. .01 D. .11 E. None

57. Divide:

.15 by .1 =

A. 15 B. 1.5 C. .15 D. .015 E. None

58. Find the percent:

90 is what per cent of 200 ?

A. 90% B. 99% C. 45% D. 49% E. None

59. In three successive shifts, "B" shift spent $34.19, $28.16, and $42.22 for meals. What was the total cost for the meals during this period?

A. $103.17.
B. $104.57.
C. $113.07.
D. $117.03

60. The Firefighter Auxiliary earned a $120.00 commission from the sale of magazine subscriptions. This amount was equal to 25% commission on all sales. How many dollars worth of magazine subscriptions did they sell?

A. $240.00 B. $480.00 C. $340.00. D. $680.00.

61. Firefighter Audet is 5 feet, 9 inches tall and Captain Hiatt is 67 inches tall. Which man is taller.

A. Firefighter Audet.
B. Captain Hiatt.
C. They are the same height.
D. None of the above.

62. Fire Engineer Simpson wishes to construct a storage box 12 feet long to keep spare equipment in. If each piece of equipment requires 4 square feet of space, how wide should he construct the box ?

A. 6 1/2 feet B. 6 feet 8 inches C. 20 feet D. 80 feet.

63. Firefighter Wendl has an average of 150 points for 12 exams. He takes 6 more exams. What would he have to average on the 6 exams to raise his average for 18 exams to 160 points.

A. 170 B. 175 C. 180 D. 190

64. Fire Engineer Wilkinson wishes to construct a storage box 12 feet long and will store 20 Firefighter groups of safety equipment (turnout coats, pants boots and breathing apparatus). If each group of safety equipment requires 4 feet of area, how wide should he construct the storage box?

A. 6 feet 8" B. 5 feet C. 20 feet. D. 30 feet.

65. If Captain Sutton can do a pre-fire plan in 10 hours and Captain Hiatt can do the same pre-fire plan in 15 hours, how many hours would it take both Captains, working simultaneously to do the pre-fire plan?

A. 5 hrs. B. 6 hrs C. 13 hrs D. 24 hrs

66. At what point will a burning substance cease giving off hot gases?

A. When it is cooled by water.
B. When it is allowed to burn freely.
C. When it is exposed to a chemical spray.
D. When it is exposed to cool air.

67. The maximum percent of oxygen in the atmosphere which can support combustion is 100%, what is the minimum per cent of oxygen that the atmosphere must contain in order to support combustion?

A. 10 % B. 12 % C. 15 % D. 16 %

68. In order for fire to continue to burn there must be fuel, oxygen, and a sufficiently high enough temperature to maintain combustion. The most common method used to extinguish fire:

A. Is the removal of the oxygen.
B. Is the removal of the fuel.
C. Is the removal of the heat.
D. Is to smother the flames of the burning material.

69. What would be the primary danger at a fire incident in a structure that contained carbon dioxide cylinders?

A. The flammability of the carbon dioxide.
B. Carbon dioxide will increase the fires intensity.
C. The carbon dioxide cylinders may explode.
D. Toxic fumes may be released from the carbon dioxide cylinders.

70. A pulley on a small electric hose winder turns clockwise. A crossed belt turns a much larger pulley for a larger hose winder. The pulley for the larger hose winder turns:

A. Counterclockwise and faster than the smaller hose winder.
B. Clockwise and faster than the smaller hose winder.
C. Counterclockwise and slower than the smaller hose winder.
D. Clockwise and slower than the smaller hose winder.

71. What is the term used to define the attraction that exist between the earth and the bodies on or near it:

A. Force B. leverage C. Vibration. D. Gravity.

72. Fire hose should not be cleaned with the use of gasoline because:

A. It would cause it to rot.
B. It would weaken the couplings.
C. It would stain it.
D. It would shrink it.

73. Any substance which offers a very high or very great resistance to the passage of electricity through it is called a/an:

A. Fuse B. Conductor C. Insulator D. Closed circuit.

74. Which of the following types of **OIL** soaked rags is most susceptible to spontaneous ignition:

A. Motor oil B. castor oil C. Linseed oil D. Fuel oil

75. What is the substance that is usually used for electrical conductors?

A. Rubber B. Glass C. Porcelain D. Copper.

76. How many blocks are there in the drawing below?

A. 2 Blocks.
B. 3 Blocks.
C. 4 blocks.
D. 5 blocks

77. How many blocks are there in the drawing below?

A. 15 Blocks.
B. 18 Blocks.
C. 19 blocks
D. 21 blocks.

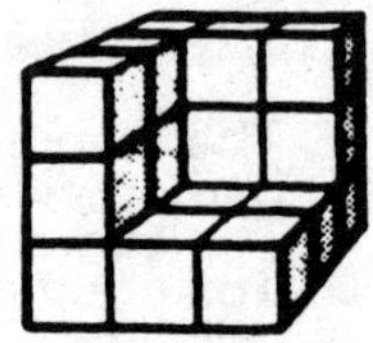

IN THE FOLLOWING DRAWINGS, FIND THE CUBE AT THE RIGHT THAT COULD BE THE ORIGINAL CUBE ROTATED INTO A DIFFERENT POSITION. CUBES MAY BE ROTATED INTO ANY POSITION AND THERE IS A DIFFERENT DESIGN ON EACH OF THE SIX SURFACES OF EACH CUBE (questions #21 - #25)

78.

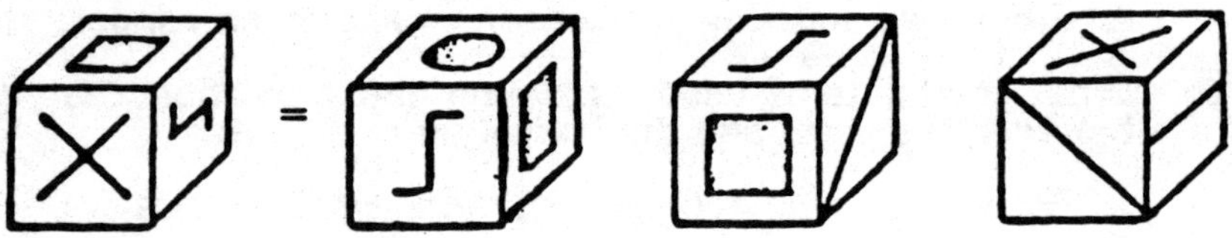

ORIGINAL A. B. C.

79.

ORIGINAL A. B. C.

80.

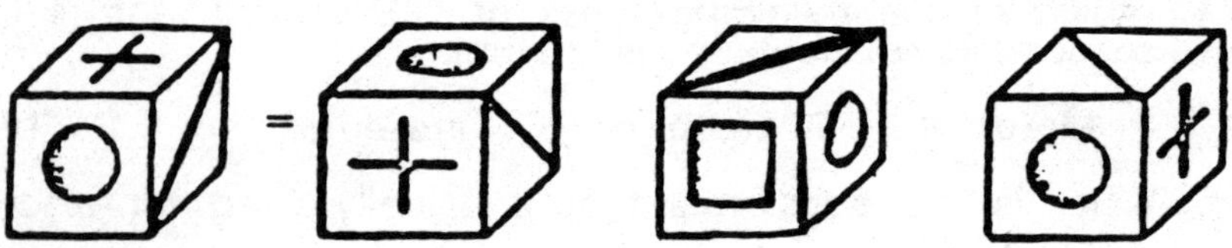

ORIGINAL A. B. C.

COMPLETE THE FOLLOWING ALPHABETICAL AND NUMERICAL PROGRESSIONS:(questions #81 - #89)

81. K M O Q __

A. R B. S C. T D. U

82. V T R P __

A. Q B. M C. N D. O

83. CC EE GG II ____

A. HH B. JJ C. KK D. LL

84. 120 124 62 64 32 36 18 22 __

A. 20 B. 22 C. 26 11 D. 28

85. 5 10 7 14 11 22 19 __

A. 17 B. 36 C. 10 D. 38

86. 3 6 11 17 26 37 50 __

A. 52 B. 50 C. 65 D. 87

87. 4 6 3 7 9 6 10 __

A. 12 B. 14 C. 16 D. 22

88. 5 10 7 14 11 22 19 __

A. 58 B. 52 C. 38 D. 32

89. 14 16 13 17 12 18 11 __

A. 18 B. 19 C. 21 D. 25

90. Of the following terms, which best describes the result of a direct blow upon a muscle?

A. Contusion B. Fracture C. Sprain D. Strain.

91. Suppose you, as a Firefighter on your way to work observe a woman at a bus bench suddenly collapse to the ground with no signs of respirations, what action would you take in this situation?

A. Call the Police Department.
B. Call the Fire Department.
C. Administer "First Aid" measures to restore breathing (artificial respiration).
D. Take no action since the woman is obviously dead.

92. Of the choices below, which best describe the symptoms of HEAT EXHAUSTION?

A. Abdominal cramps, red skin, and a large amount of perspiration.
B. Strong rapid pulse, fever with dry skin.
C. Weak pulse, low temperature, pale clammy skin.
D. Headache, red face, followed by unconsciousness

93. Of the choices below, which one describes the main reason for administering artificial respiration?

A. To force air into the patients lungs.
B. To pump the patients heart.
C. To circulate the patients blood.
D. To keep the patient warm.

CHOOSE THE CORRECT SPELLING FOR THE WORDS IN THE FOLLOWING:(questions #94 - #98)

94.
A. mezzine.
B. mezanine.
C. mezzanine.
D. none of the above.

95.
A. millameter.
B. millemmetor.
C. millemetre.
D. none of the above.

96.
A. multiple.
B. multaple.
C. multipel.
D. none of the above.

97.
A. lattitude.
B. latatude.
C. latitude.
D. none of the above.

98.
A. legitimate.
B. legitamate.
C. lagitimate.
D. none of the above.

IN THE FOLLOWING SELECT THE SENTENCE THAT HAS AN ERROR IN GRAMMAR, USAGE, OR PUNCTUATION:(questions #99 & #100)

99.

A. I feel badly because of the error.
B. The proposal sounds hopeful.
C. He was beside me in the stadium.
D. You look attractive to me.

100.

A. To understand her is to appreciate her.
B. This affair is totally between you and me.
C. She was an expert, a teacher, and could she speak!
D. I should like to have this opportunity ten years ago.

PRACTICE EXAM #10
ANSWERS

1. A
2. B
3. C
4. D
5. A
6. C
7. C
8. C
9. D
10. D
11. C
12. D
13. B
14. A
15. C
16. B
17. A
18. A
19. B
20. C
21. C
22. B
23. D
24. A
25. C
26. C
27. D
28. C
29. B
30. C
31. A
32. C
33. D
34. D
35. B
36. D
37. C
38. B
39. A
40. B
41. C
42. D
43. B
44. A
45. A
46. C
47. C
48. A
49. D
50. B
51. C
52. B
53. B
54. C
55. E
56. C
57. B
58. C
59. B
60. B
61. A
62. B
63. C
64. A
65. B
66. A
67. B
68. C
69. C
70. C
71. D
72. A
73. C
74. C
75. D
76. C
77. C
78. C
79. C
80. B
81. B
82. C
83. C
84. C
85. D
86. C
87. A
88. D
89. B
90. A
91. C
92. C
93. A
94. C
95. D
96. A
97. C
98. A
99. A
100. C

PRACTICE EXAM ANSWER FORM

1	A B C D E	26	A B C D E	51	A B C D E	76	A B C D E
2	A B C D E	27	A B C D E	52	A B C D E	77	A B C D E
3	A B C D E	28	A B C D E	53	A B C D E	78	A B C D E
4	A B C D E	29	A B C D E	54	A B C D E	79	A B C D E
5	A B C D E	30	A B C D E	55	A B C D E	80	A B C D E
6	A B C D E	31	A B C D E	56	A B C D E	81	A B C D E
7	A B C D E	32	A B C D E	57	A B C D E	82	A B C D E
8	A B C D E	33	A B C D E	58	A B C D E	83	A B C D E
9	A B C D E	34	A B C D E	59	A B C D E	84	A B C D E
10	A B C D E	35	A B C D E	60	A B C D E	85	A B C D E
11	A B C D E	36	A B C D E	61	A B C D E	86	A B C D E
12	A B C D E	37	A B C D E	62	A B C D E	87	A B C D E
13	A B C D E	38	A B C D E	63	A B C D E	88	A B C D E
14	A B C D E	39	A B C D E	64	A B C D E	89	A B C D E
15	A B C D E	40	A B C D E	65	A B C D E	90	A B C D E
16	A B C D E	41	A B C D E	66	A B C D E	91	A B C D E
17	A B C D E	42	A B C D E	67	A B C D E	92	A B C D E
18	A B C D E	43	A B C D E	68	A B C D E	93	A B C D E
19	A B C D E	44	A B C D E	69	A B C D E	94	A B C D E
20	A B C D E	45	A B C D E	70	A B C D E	95	A B C D E
21	A B C D E	46	A B C D E	71	A B C D E	96	A B C D E
22	A B C D E	47	A B C D E	72	A B C D E	97	A B C D E
23	A B C D E	48	A B C D E	73	A B C D E	98	A B C D E
24	A B C D E	49	A B C D E	74	A B C D E	99	A B C D E
25	A B C D E	50	A B C D E	75	A B C D E	100	A B C D E